The Stellar Lifecycle

Amateur Astrophotography
Tells The
Story Of The Stars

A PRACTICAL GUIDE TO IMAGING & UNDERSTANDING
THE STELLAR LIFECYCLE

ISBN: 978-1-7266-4709-0

DEDICATION

This book is dedicated to my wife Abigail, who has allowed and even encouraged me to spend many hours outside in the dark on my own (!), welcomed me back into a warm bed with freezing feet, and tolerated my collection of a bewildering array of gadgets and equipment. All my love, Jerry.

ACKNOWLEDGMENTS

My journey into the universe began as a teenager with the encouragement of a stranger. The impact of public outreach on the lives of those we touch can't be underestimated, particularly when those engaged are the next generation of scientists, engineers, explorers, artists and dreamers. Thanks to Stuart's energy and passion, I have enjoyed over 30 years of astronomy as an amateur and went on to become a professional engineer.

CONTENTS

1 LIGHT, THE UNIVERSE AND EVERYTHING

In writing this book I wanted to share my lifelong passion for astrophotography and demonstrate to those tempted to embark on their own journey that it is entirely possible to start from nowhere and achieve some amazing things; all with a little encouragement and a lot of perseverance. I hope that you enjoy my images and they inspire you to take your own. The "how to" sections in Appendices I to III are aimed at getting you started and then moving you on to more complex imaging techniques.

Rather than just presenting a collection of my images, I wanted to demonstrate how an amateur can capture the full lifecycle of a star, from its birth through to its death, using relatively modest equipment. At this early juncture, I must say that I am not a physicist - certainly not an astrophysicist, but I've done my best to explain the stellar lifecycle as accurately as I can! And I beg forgiveness for any inaccuracies I might impart in my amateur enthusiasm.

"In the beginning...there was light". These immortal words penned in our early recorded history reflect man's yearning to understand where everything came from and his place in the universe. Our eyes have always been drawn up to the stars, although perhaps less so today in the light-drenched skies of our towns and cities, admiring the beauty and majesty beyond our own existence. Growing up in the Lake District National Park, in the north of England, I was fortunate to have very dark skies, albeit they were quite often filled with cloud.

But what are we looking at when we see the stars shining down on us? What are those cloud-like nebulae, made famous in glorious colour photographs by observatories such as the Royal Observatory Edinburgh, the Hubble Space Telescope, Spitzer and others? And is it possible to take images of your own, to capture the beauty of the universe?

So how hard can it be, taking pictures of bright objects shining against the inky blackness of the night sky? Well, going back over thirty years to a time when publicly available computers included the IBM PS/2 (not to be confused by younger readers with the PlayStation 2!) and 5 years before the sale of any digital SLR cameras, it's fair to say that taking pictures of the night sky was a very hit and miss affair as a novice. Mainly miss. I think my success rate was less than 5%, with a roll of 36 exposures only yielding 1 or 2 useful pictures! But move forward in time to the 21st century and amateur astronomers are spoiled by an array of relatively cheap digital cameras, star-tracking devices and powerful software packages that make it quite simple to produce impressive images. Dare I say it, images that can rival or even exceed those produced by the major international observatories of the 20th century, although it's a long way for me to reach that lofty pinnacle. Of course, some things have not improved over that time, with light pollution now becoming a real threat to future generations actually being able to see the night sky.

Before turning to the science of the stellar lifecycle and some images, this first chapter will introduce some of the challenges of astrophotography and discuss the root cause of our difficulties. Ignoring our own star, the Sun, the brightest star in our night sky is Sirius in the constellation Canis Major, shining very brightly on the southern horizon from the UK at magnitude -1.46. It's readily visible from a light polluted town and is a lovely sight with the naked eye from the countryside, blazing brilliantly below the recognisable constellation of Orion. But how does the brightness of this star compare to things that we would quite easily take pictures of in the daytime, perhaps with a smartphone or DSLR camera?

1

Paul Schlyter's *Radiometry and Photometry In Astronomy* cites the illumination of various natural light sources as:

Object	Illuminance (Lux)	Stellar magnitude
Indirect sunlight	10,000 to 25,000	-24 to -25
An overcast day	1000	-21
Full moon overhead	0.267	-12.5
Sirius (the brightest star)	0.00001	-1.46

Without going into too much detail, the apparent brightness of a star to us here on earth (noted as stellar magnitude above) uses a logarithmic scale. So, each change in magnitude of 1 unit equates to a change in brightness of approximately 2.5 times. Or more accurately, a difference of 5 magnitudes corresponds to a brightness factor of 100.

"The stars glowed like jewels thrown across a black velvet scarf"
Oliver Andressan

The constellation of Orion (centre) and Sirius, the brightest star in the sky, bottom left near the horizon

The data on page 2 show that even the brightest star in the night sky is eight orders of magnitude fainter than the light levels available to a camera on a typically cloudy day. In other words, the light hitting the camera sensor is 100 million times fainter! Little wonder then, that instead of exposures of hundredths or thousandths of a second, that are typical in normal terrestrial photography, astrophotos are often measured in seconds, minutes or even hours - per exposure. And herein lies one of the problems.

Tracking The Stars

At the equator, the earth spins on its axis at a speed of about 1000 miles an hour. Why can't you feel that? Simply because you and the atmosphere are all spinning with the earth at the same relative speed, so it feels like we're not moving. A long-exposure image of the night sky, taken with a camera on a fixed tripod, quickly shows how fast the earth is turning, with star trails revealing the apparent motion of the stars across the sky. This image shows 70 minutes of rotation.

Circumpolar star trails, revealing the apparent motion of the stars around the north celestial pole star Polaris

As the earth rotates once on its axis every 23 hours 56 minutes and 4 seconds, we can say that its rotational or angular speed is 360 degrees per "day", where the day is a little bit less than 24 hours. Degrees are very large units, which are subdivided into minutes and then seconds, with 60 per subdivision. So, with a quick calculation we can see that the angular speed is:

$$\text{angular speed} = 360\text{deg} / 23\text{h}56\text{m}4\text{s} = (360*60*60) \text{ arc-seconds} / (23*60*60)+(56*60)+4 \text{ seconds}$$
$$\text{angular speed} = 1{,}296{,}000 \text{ arc-seconds} / 86{,}164 \text{ seconds}$$
$$\text{angular speed} = 15.04 \text{ arc-seconds/second}$$

To put the angular measurement of one arc-second into context, it is the angle subtended by a pound coin at a distance of approximately 4km. Which is pretty small. But of course, in astronomy we are usually looking at objects in the sky that have a very small *apparent* size. So, we often zoom in with long focal length lenses to magnify the view. And doing so, we come across the problem of the movement of that object across the sky in a short period of time.

Generally speaking, when using a camera on a fixed tripod, the maximum exposure time of an image that can be taken before stars begin to trail on the photo, is inversely proportional to the focal length of the lens. The longer the focal length of the lens, the shorter the exposure time. A typical rule of thumb is:

$$\text{maximum exposure time} = 500 / \text{focal length of lens (in mm)}$$

For a nice wide-angle lens, say 35mm the maximum exposure time on a fixed tripod is about 14 seconds. Whereas, for a 200mm telephoto lens, this rapidly decreases to 2.5 seconds. In practice we can stretch these figures a little, perhaps even doubling them if one is a little less fussy about having perfectly round stars. However, this is sufficient to demonstrate that if you want to take images of many minutes, a device that counteracts the rotation of the earth is vital. We refer to this as a tracking device of some kind.

Focusing

The second challenge presented by the faint light source is accurate focusing. The autofocus mode on most cameras will struggle to function with the very faint signal detected in the instantaneous focusing mode. And whilst focusing the lens at infinity was simple with non-digital cameras, digital cameras often go beyond this focus point and then come back in a little due to the way the focus system works. So, we can no longer just twist the lens to its fullest extent and be certain that we are focused at infinity.

There are a few simple techniques to help though. If the camera has a "live view" mode, display this whilst focusing on a bright star and make tiny adjustments to the focus manually until you are happy that the star is as sharp as possible. Better still, there are a couple of simple masks that you can buy or even make yourself to remove any doubt about whether your lens is in focus. I prefer to use the Bhatinov mask, but a Hartmann mask works almost as well.

The mask is placed in front of the lens whilst focusing and works by introducing a diffraction pattern that changes as you approach the focal point.

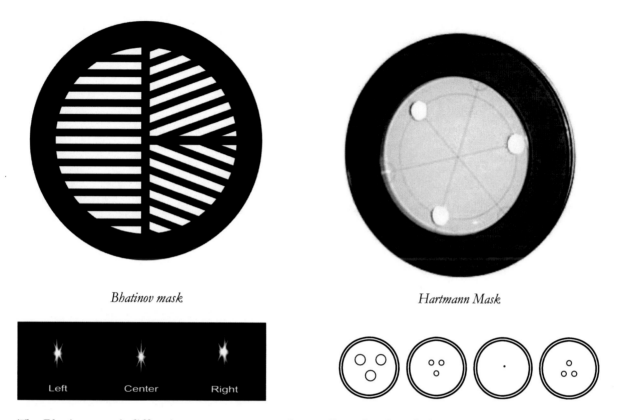

Bhatinov mask

Hartmann Mask

The Bhatinov mask diffraction pattern presents three spikes, showing obvious misalignment when the star is inside or outside of the focal point. Alternatively, the Hartmann mask diffraction pattern provides three images of the same star, which get closer together as the focal point is approached, ultimately blending into one star at optimum focus. Both tools are fool proof ways of focusing and remove any subjectivity from the process.

These masks can be purchased commercially, but it's just as easy to make one yourself with a piece of paper/card, a craft knife and if possible a laminator. The pattern of the Bhatinov mask varies with the size and focal length of the lens. But there are plenty of pattern generator tools on the internet, which you can input your lens or telescope details into.

Quality of Optics

A third issue that becomes quite quickly evident when imaging the stars is the quality of the optics in the lens or telescope. Because the stars are point sources rather than extended objects, they should be nice tight round objects spread over only a few pixels. However, many lenses suffer from imperfections such as (i) chromatic aberration, the dispersion of different wavelengths of light that often appears as blue or violet halos around stars; (ii) spherical aberration, where the light is not evenly focused across the whole lens or mirror surface; (iii) coma, where stars at the edge of the field of view are distorted into tadpole shapes.

For the beginner, all of these issues can be ignored. Better to commence with basic equipment and learn the ropes, investing in more expensive lenses and telescopes at a later stage. However, the cost of high-quality optics has significantly reduced over the past twenty years, with a good apochromatic telescope available for under £500 now.

High Dynamic Range (HDR)

Another challenge that is more obvious in objects such as the Andromeda Galaxy (M31) or the Orion Nebula (M42), is the wide range of brightness in the same object. The core of these objects is very bright in comparison to the outer areas. So, a range of exposures may be required to capture the bright portions of the image, without over-exposing them, as well as the much fainter outer areas. The principle of HDR processing is now common in terrestrial photography, with camera phones even compensating for different brightness ranges in the image automatically. Of course, in astrophotography this means a little more work, capturing short, medium and long exposure images of the same object and then combining them later using software such as Photoshop or other imaging packages.

Low Levels of Light

Returning to the issue identified at the start of this chapter, it can be said that the most obvious challenge is gathering enough light to see something spectacular. Although the stars are enormous power-houses of energy and radiation, they are so far away from us that the light they deliver to earth is very little. And when we are looking to image glowing areas of gas or dark patches of dust, not stars themselves, the challenge becomes even bigger.

There are two fundamental properties of the equipment that are important in terms of capturing the faintest signals as quickly as possible, and three basic techniques taking the images that also help. Not all of these aspects need to be considered at the same time, but to get the best image this is often the case. They are presented here in relative order of importance (in my opinion)!

(1)   Large, fast lenses/telescopes (equipment properties)

The amount of light gathered by the lens or telescope is directly proportional to the square of its diameter. For a given diameter of lens, the shorter the focal ratio the faster the lens. So, it's a combination of the diameter and the focal ratio that enables the most light to be captured in the shortest time. When talking about camera lenses, a "fast lens" would be something like f/2.8. But remember, the diameter is critical, the bigger the better. I have three telescopes that I use for imaging, all of which are optimised in one way or another for photographic use, as shown in the following table.

Telescope	Type	Diameter	Focal Ratio	Focal length
Skywatcher Maksutov-Newtonian MN190	Mirror	190mm (sub-8")	f/5.2	1000mm
Takahashi E130D	Mirror	130mm (5.1")	f/3.3	430mm
TS INED 70mm apochromatic	Lens	70mm (2.75")	f/6	420mm
(fitted with 0.8x reducer)		70mm	f/4.8	336mm

The benefit of having a selection of telescopes/lenses and a variety of cameras is that they can be mixed and matched to the type of object being imaged. For smaller objects a longer focal length is required, whilst for large objects spread out over several degrees a short focal length is required. Although reducers and extenders can help broaden the range of a given instrument, there is often no substitute for the inherent properties of the telescope in terms of diameter and f/ratio.

(2)   Sensitive cameras (equipment properties)

With respect to digital cameras, there are two basic types of sensor that are used to capture the light signal (photons). One is called a Charge Coupled Device (CCD) and the other is a Complementary Metal-Oxide Semiconductor (CMOS). Each have their own benefits and disadvantages, and the market has evolved hugely over the past decade. There are various characteristics that are important in the selection of a camera, but at its simplest it can be said that dedicated astrophotography cameras outperform standards digital SLRs, as one would expect due to their specific design features. However, many astrophotographers choose to use DSLRs, myself included, due to their simplicity of operation. And some fantastic images can be taken with nothing more than a DSLR and a standard camera lens. In addition to my trusty Canon 300D, which was one of the first DSLR cameras produced that was really suited to astrophotography, I also have two dedicated astronomy cameras.

Camera	Sensor	Megapixels	Chip-Size	Chip-Cooling		Quantum-Efficiency
QHY23M	CCD	9MP	12.5 x 10mm	-45 deg C	(Δ)	60-70%
QHY163M	CMOS	16MP	17.7 x 13.4mm	-40 deg C	(Δ)	Unknown but >50%
Canon 300D	CMOS	6MP	22.7 x 15.1mm	None		Unknown but <40%

One of the most interesting characteristics of these dedicated astronomy cameras is the very high quantum efficiency. Simply put, this is the proportion of photons that are captured by the sensor as a signal, as a percentage of the actual number of photons that hit it. Some Back-Illuminated CMOS chips are now >90% efficient! A more typical sensitivity level though for CCD and CMOS cameras is between 50-70% QE, with the response varying across the spectrum with some chips being more responsive at the important wavelength of hydrogen-alpha emission.

By comparison, the original Canon 300D SLR camera had two disadvantages. Firstly, the quantum efficiency of the sensor was of the order of 30-40%. And secondly, the presence of a filter on the front of the chip, which corrects for the right daytime colour balance, blocks a large proportion of the red hydrogen-alpha emission line - probably the most important emission line in astrophotography! My camera has been modified by having this filter removed, to ensure that the chip receives as much light from the nebulae as possible, but this is far from essential for the beginner.

Another major benefit of CCDs is the ability to combine pixels together, called binning. Binning 2x2 creates a super-pixel that is 4 times more sensitive than the single pixel. Binning 4x4 is theoretically 16 times more sensitive. In reality the increased sensitivity isn't linear, with 2x2 yielding more like 3x sensitivity and 4x4 yielding around 10x. Although binning decreases the resolution, because there are now less pixels covering the same surface area of the optical field of view, the boost to sensitivity can be very useful when imaging faint objects.

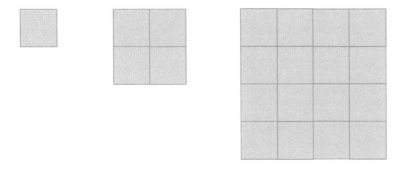

Pixel binning utilises the hardware in the CCD chip in a clever way to boost the sensitivity of the camera. Other benefits include (i) improving the dynamic range of the super-pixel, meaning longer exposures can be taken without saturating the stars and thereby losing colour fidelity; and (ii) improving the signal:noise ratio inherent in the chip since the super-pixel is read out and converted in the amplifier only once.

The diagram above shows a single pixel, a large 2x2 pixel and a super-pixel array of 4x4.

(3)   Long exposures (imaging technique)

The most effective way of capturing the faint signal is to take long exposures. Excluding the sun, moon and planets, all other forms of astrophotography require exposures of many seconds, or more typically minutes. It's difficult to state exactly how long each exposure should be, as it depends on several factors including:

- The brightness of the object - for example the Orion nebula (M42) can be captured in a matter of seconds, compared to the elusive Spaghetti nebula which takes many minutes per exposure to begin to show on the image. And of course, the brightness range of the object influences the choice of exposure time as already discussed (HDR);
- The amount of background light in the sky, caused by poor seeing conditions, the presence of moonlight, levels of light pollution, or how far below the horizon the Sun is (particularly during summer time in Britain);
- Whether any filters are being used, such as narrowband filters like H-Alpha, Sulfur-ii or Oxygen-iii. Generally speaking, when filters are being used the exposure will need to be longer than without filters.

As a general guide, the following table gives further detail of typical exposures that could be used on various objects. Of course, trial and error are part of the joy of astrophotography!

When using narrowband filters, it is typical to take individual exposures of between 300s and 1200s. One of the benefits of this approach is that the contrast between the nebula and the general sky background is much better. Although the accuracy of tracking becomes extremely important over such extended durations.

One drawback of CMOS cameras is that they can suffer from amp glow around the edges of the chip, although some manufacturers such as QHY are beginning to combat this with clever technical innovations. One of the drawbacks of this aspect of CMOS chips is that it can become very noticeable on exposures over 300 seconds duration. This is particularly the case in my old Canon 300D and no amount of image calibration or post-processing can fully remove the artefacts. This is where cooled CCD cameras reign supreme, taking long exposures without this problem.

Object	Suggested Exposure	Comments	
Double Cluster (Perseus) NGC884 & NGC869	300s	There is no gas or dust in this object, so all we are imaging is starlight. In order to keep the star colour nicely saturated and not overexposed, each exposure should be between 60-300 seconds. With a DSLR camera the ISO setting should be no more than 400.	
Andromeda Galaxy (M31)	30-60s 300-600s	A range of exposures are needed in order to capture the faint outer arms of the galaxy and the bright inner core. Short exposures of 30-60s should enable the core not to be massively over-exposed. And longer exposures of 300-600s will capture the full extent of the outer galactic arms. The two sets of exposures will need to be processed independently and then combined as an HDR image. Details of how to do this are given in Appendix III.	
Great Orion nebula (M42)	10s 30-60s 300-600s	A range of exposures are needed in order to capture the very high brightness range of this object. Long exposures will reveal the outer faint edges of the nebula, whilst short exposures of only a few seconds will capture the bright stars at the centre of the nebula and the very bright core of gas in the central regions. Use HDR image combination techniques to blend these together.	

Typical exposures for a range of deep-sky objects

(4)   Multiple exposures

The benefits of taking multiple exposures and combining them to produce a single image cannot be over-emphasised. The basic intent is to improve the signal-to-noise ratio (SNR) in the final image, thus enabling the image to be "stretched" whilst maintaining image quality. But what is meant by signal and noise?

Signal is the genuine photon arriving at the camera chip from the object being photographed. All things being equal, such as accuracy of tracking, alignment of camera and optics, atmospheric conditions, etc. the signal will always be in the same place on the chip. But in the context of astrophotography, image noise is unwanted data in a pixel that is not true photonic signal. It comes from various sources, but can be simplified to the following key aspects; (i) the Poisson noise associated with the quantum nature of light, which is random; (ii) thermal noise generated by the image sensor itself, sometimes called the dark current of the pixels; (iii) read noise generated by the electronics in the sensor.

The second and third sources of noise are predictable characteristics of the camera sensor and these can be removed using calibration techniques called "dark" and "bias" subtraction. The first source of noise can be minimised by taking multiple exposures and combining them, or stacking them, using clever mathematical techniques. At its simplest, if the data in each image is combined using an "average" function then the signal remains constant but the noise, which is random, decreases. Hence the signal:noise ratio increases and the image becomes purer. But how many images should be stacked?

The signal:noise ratio is proportional to the square root of the signal. So, to double the signal:noise ratio, it requires four times the signal. To double it again, or make it four times better, we would need to take 16 exposures. Clearly, there is a law of diminishing returns with more and more exposures required to significantly change the SNR. In practice, the number of exposures taken are often limited by how much time is available to image a single object, or perhaps by the patience and perseverance of the photographer. The illustration below demonstrates how an image can be stretched to reveal the fainter areas much more effectively with even a small stack of images. Note that both images have had exactly the same stretch applied.

A single 5-minute exposure of M101

An average stack of 6 x 5-minute exposures

(5)   Narrowband filters

The third technique that adds huge value is narrowband imaging. But what is it? Simply put, a narrowband filter is a rejection filter, in that it only allows a narrow portion of the electromagnetic spectrum to pass through the filter at a specific wavelength of light. All other wavelengths are blocked or rejected. Why do we want to do this, and what are the wavelengths of light that we want to capture?

The most common element in the universe is hydrogen. And when hydrogen is ionised by electromagnetic radiation it most commonly loses an electron and becomes unstable. The nucleus naturally wants to recombine with an electron, which is usually at a higher energy level. The newly created hydrogen atom now wants to stabilise itself, so the electron drops down to the base level. About 50% of the time, the electron jumps from the third quantum level to the ground level and thereby releases a photon at 656.281nm, in the mid-red portion of the spectrum. Another slightly less common emission line for hydrogen is hydrogen-beta, emitted when the electron jumps from the fourth energy level to the second, emitting an electron at a different wavelength of 486.136nm, in the blue-green portion of the spectrum.

Other common emission lines important to astrophotographers are doubly-ionised oxygen (O-iii), which emits primarily at 500.7nm (green-blue); and singly-ionised sulfur (S-ii), which emits at 671-673nm (dark-red). The standard colour palette used by the Hubble Space Telescope is S-H-O, with narrowband images mapped to red, green and blue channels respectively to create a colour image. Clearly, this is a false colour image, made from red, red and green-blue light. Sometimes, these filters can be used just to enhance contrast in a standard broadband colour image.

When using a monochrome camera, a broadband image is created using red, green and blue filters, with an optional clear luminance channel sometimes. Broadband imaging is required for objects like galaxies and stars which emit light across the full spectrum and are not limited to narrow windows surrounding key wavelengths like those mentioned above.

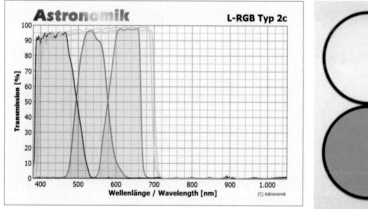 

LRGB filters are used to create full-colour broadband images across the whole of the visible spectrum

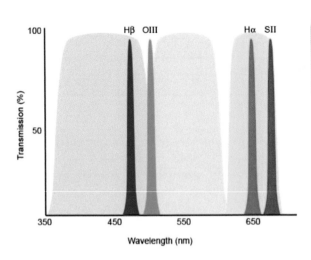

The number suffix on the filter denotes the width of the bandpass allowed through the filter in nanometers. The H-α filter below allows light in the wavelength of 653nm to 660nm to pass. The narrower the bandpass, the higher the contrast it provides, although the image is darker due to passing less light.

H-α 7   H-β 8.5   O III 8.5   S II 8

Narrowband filters can be used to create false-colour images like those produced by the Hubble Space Telescope

One of the benefits of narrowband imaging is that it removes a huge amount of light pollution and general background glow from the sky. The narrowband filters reject the emission of sodium, mercury, tungsten and other artificial lights that commonly spoil our view of the night sky in towns and cities. Thus, the city dweller can now take stunning deep-sky images that were impossible for the amateur only 20 years ago. Another benefit of using these filters is being able to take deep-sky images even when there is moonlight brightening the sky.

Although I do produce some images using only narrowband filters, I prefer to combine data using both broadband and narrowband filters to produce a single image that really "pops" with detail. The narrowband data greatly enhances contrast in nebulae and when mapped carefully in a colour image, can add a more three-dimensional aspect, whilst the broadband data provides the natural colours.

In summary, the main challenges faced in deep-sky astrophotography are:

- Tracking the stars as the earth rotates, often for many hours
- Focusing on very faint objects
- Overcoming optical imperfections
- Coping with a high range of brightness in a single image (HDR composition)
- Recording as much light as possible

Overcoming these challenges is part of the fun, and doing so enables the amateur astronomer to capture the full stellar lifecycle using their own equipment and perseverance! The next five chapters present the physics of the stars, revealed in amateur images taken from the dark skies of England's Lake District National Park, using relatively modest equipment.

I hope you enjoy it!

2 A STAR IS BORN

"What is a star? Why does it shine? Where do stars come from? Do they last forever?" These are all questions I've been asked when teaching astronomy club at my local primary school, and are probably questions we've all thought about when we were young. If we travelled back in time only a hundred years we wouldn't have the answer to these questions, which makes it fascinating when you consider how much our understanding of the universe has grown over the recent past. It was only in 1990 that scientists confirmed Bart Bok's 1940s theory of star formation within "Bok globules"! And our understanding grows all the time, often with the help of amateur astronomers.

A brilliant example of this is the discovery of the Integrated Flux Nebula in 2004 by American amateur astrophotographer Steve Mandel, who took very long exposures looking out of the galactic plane and found extremely faint nebulae that are illuminated by the combined starlight of our galaxy, rather than being directly illuminated by nearby stars. It should be noted that the presence of interstellar dust or galactic cirrus was catalogued around 1965 by Lynds but was not fully understood; and it was Mandel's amateur images that led to a professional collaboration with Kitt Peak National Observatory to photograph and catalogue the Integrated Flux Nebulae. But I digress! The aim of this chapter is to briefly discuss the main stages of the stellar lifecycle and look at where it all begins. Although these stages are not scientifically classified in this way, I find it a good generalisation that helps explain the physical processes and categorises the types of images we might try to take.

Stage 1 - Protostar Formation (Where it all begins) - Giant Molecular Clouds
Stage 2 - A Star Is Born (The Young Ones) - Hot young stars embedded in their birth nebula; multiple star systems
Stage 3 - Growing Up (Teenage years to adulthood) - Open Clusters; Globular Clusters; Galaxies
Stage 4 - Old Age (Middle-aged spread) - Planetary nebulae; other nebulae
Stage 5 - Death & Rebirth (The circle of life) - Supernovae eruptions; supernova remnant nebulae and degenerate stars

Interestingly, you can image all stages of the lifecycle in one picture if you're lucky...but we'll come to that later (p50).

So, what is actually happening in the very first stages of stellar evolution? First let's consider the general nature of matter distributed throughout the universe, ignoring discrete objects such as stars, planets, etc. Intergalactic space is on average very empty, being generally free of dust and gas in comparison to the space within galaxies, although of course it's not totally empty. (I'm also ignoring the tricky topic of dark matter here...) The density profile of the space within galaxies varies significantly, with some "empty" areas and some areas of higher density gas and dust. The gas is in atomic form, in molecular form and some is ionised, which is sometimes referred to as the three-phase model. Often, the gas we image is a combination of all three states.

But the area of most interest in terms of star formation is the molecular hydrogen, which clumps together to form molecular clouds or Giant Molecular Clouds up to several hundred light years in diameter. The density of these clouds can typically be a thousand particles per cubic centimetre, although the density profile varies significantly within the cloud. This is in comparison to a particle density of 0.1 to 1 particle per cc in the "empty" volumes of space.

"For my part I know nothing with any certainty, but the sight of the stars makes me dream"
Vincent Van Gogh

A wide-field view of part of the Giant Molecular Clouds in the Orion arm of our galaxy

The image has higher resolution data added in to reveal more of the Great Orion Nebula (M42/43) and separately of the Horsehead and Flame Nebulae as follows.

Camera/lens: QHY163M with a 90mm telephoto lens (wide-field image)
Exposures: H-alpha (11x300seconds); Luminance (6x300seconds bin2); red, green and blue (8x180s bin2) each
Camera/scope: QHY23M with TSINED70mm & 0.8x reducer (336mm f/4.8)
Exposures: H-alpha (7x600seconds; 10x300seconds; 8x60seconds; 12x10seconds all bin2) for M42/43 Nebulae
Camera/scope: QHY23M with Takahashi E130D (430mm f/3.3)
Exposures: H-alpha (6x600seconds for a 2-panel mosaic) for B33 Horsehead Nebula & Flame Nebula

There are several interesting techniques in the image on page 14, including the use of narrowband data to enhance contrast and increase the detail of the nebulae; the use of High Dynamic Range processing to suppress the brightness of the core of the main M42 nebula; the construction of a mosaic of 2 panels for the Horsehead/Flame to encompass the entire area at a high resolution; and the use of multiple camera and lens/telescope combinations that are integrated to give a wide-field image with additional detail in certain areas. Due to the poor image scale in the wide-field view, it is useful to look at the higher resolution images separately, which we'll do in the next three images.

Note, image scale is described in arc-seconds per pixel. Under-sampling an image gives blocky pixelated stars and a low level of resolution. A good image scale is between 0.5 and 3 arc-seconds/pixel and is calculated as *image scale = 206.3 * (pixel size in microns) / (focal length in mm)*. So smaller focal length lenses give bigger (poorer) image scales unless you have tiny pixels in the sensor.

"Deep into that darkness peering, long I stood there, wondering, fearing, doubting, dreaming dreams no mortal ever dared to dream before"
Edgar Allan Poe

The Horsehead (B33/IC434) and Flame (NGC2024) nebulae, part of the Orion B GMC (hydrogen-alpha emission)

"I looked at the stars, and considered how awful it would be for a man to turn his face up to them as he froze to death, and see no help or pity in all the glittering multitude"
Charles Dickens

The Great Orion nebula (M42/M43), part of the Orion A Giant Molecular Cloud (hydrogen-alpha)

The Orion Giant Molecular Cloud (GMC) is around 1,400 light years away and hundreds of light years across. Often embedded within these GMCs are found smaller areas of even higher density gas and dust, referred to as Bok Globules. And it is this fundamental property of higher density, coupled with very low temperatures that gives rise to the birth of stars. The Orion Nebula itself (M42) is an H-II region, a later stage of the development of a GMC, where the original atomic hydrogen has been ionised by the hot young stars birthed in the nebula. More of that later though.

I find the black and white single wavelength images like these to be visually very striking. But it is also nice to see what the area looks like in full colour by combining the h-alpha data with natural colour (red, green & blue - RGB) data.

"Let both sides seek to invoke the wonders of science instead of its terrors. Together let us explore the stars..."
John F Kennedy

The Great Orion Nebula, combining broadband colour and narrowband h-alpha data

Camera/scope: QHY23M with TSINED70mm & 0.8x reducer (336mm f/4.8)
Exposures: H-alpha (7x600seconds; 10x300seconds; 8x60seconds; 12x10seconds all bin2)
Camera/scope: QHY163M with Takahashi E130D (430mm f/3.3)
Exposures: RGB (3x180seconds per filter bin1; 12x10seconds per filter bin1)

The best analogy for Bok Globules is that of a cocoon, a cold sheltered region of space that protects the atomic hydrogen within from the harmful radiation from other nearby stars. Over many millions of years, the GMC begins to contract under its own gravity, probably influenced by gravitational interactions with nearby massive objects or as a result of shockwaves from exploding stars (supernovae). And due to density fluctuations within the cloud, clumps begin to form, often associated with higher proportions of dust that shelter the hydrogen gas. Let's look at some more examples of Bok Globules in the next two images.

"The sky is filled with stars, invisible by day"
Henry Wadsworth Longfellow, c.1880

NGC281 is an H-II region with several embedded dark Bok Globules

Camera/scope: QHY23M with Takahashi E130D (430mm f/3.3)
Exposures: Sulfur-ii, H-alpha, Oxygen-iii (6x600seconds per filter all bin1)
Camera/scope: QHY163M with Skywatcher MN190 (1000mm f/5.2)
Exposures: LRGB (22x180seconds Luminance; 20x120seconds per colour filter all bin1)

"It is not in the stars to hold our destiny, but in ourselves"
William Shakespeare

*The Eagle Nebula, made particularly famous by the Hubble Space Telescope's
"Pillars of Creation" image, revealing star formation deep inside the dusty pillars*

Camera/scope: QHY23M with TSINED70mm & 0.8x reducer (336mm f/4.8)
Exposures: Sulfur-ii, H-alpha, Oxygen-iii (3x300seconds; 5x300seconds; 4x300seconds all bin2)
Camera/scope: QHY163M with Skywatcher MN190 (1000mm f/5.2)
Exposures: H-alpha & RGB (3x180seconds; 4x300seconds per colour filter all bin2)

3 "YOUR OWN EYES SAW EVEN THE EMBRYO OF ME"

So how does the star actually form within this cocoon? As the hydrogen cloud collapses it becomes denser and hotter, directly radiating this energy away for a time and then using the dust within the cloud as a secondary stage of heat removal, continuing the collapse further. Other complex processes assist the ongoing cloud collapse and increase in pressure, to the point that the internal temperature of the gas at the core is hot enough to withstand the gravitational pressure of the cloud. And an embryonic protostar is born. This process does not usually occur in isolation, with irregular clumps of gas within the cloud each forming their own protostar. But at this stage, the protostar is not yet hot enough or dense enough to trigger nuclear fusion.

The protostar continues to accrete more gas as it is drawn in towards the new massive object in the cloud over a timescale of a few million years, building the pressure and increasing the internal temperature at the core until nuclear fusion is triggered. Fusion is the combination of two atomic nuclei to form a new element, such as 2 hydrogen atoms to form a helium atom. Initially, this is likely to be deuterium-fusion as this occurs at the lowest temperature point of around 1 million Kelvin. The deuterium fusion process is relatively gentle compared to standard hydrogen-fusion, enabling the new star to continue to accrete more mass from its circumstellar disk. But eventually the deuterium fusion process is overtaken in the core by the fusion of hydrogen as the inner temperature rises to about 10 million Kelvin, pushing the deuterium-fusion process into the outer layers of the protostar until its limited supply of deuterium is exhausted. Eventually the process of accreting more mass into the star finishes as the outward radiative pressure prevents any further in-fall of material, driven by the more violent hydrogen fusion process. And the protostar is now considered to be a young but fully formed star.

Scientists using tools like NASA's Chandra X-ray observatory and the Spitzer Space Telescope continue to study opaque dusty areas that can't be penetrated by optical telescopes, revealing very young stars and protostars that are invisible to amateurs. In 2014, a study of NGC2024 (the Flame Nebula in Orion) revealed many stars under 200,000 years of age hidden inside the dark central portion of the nebula.

Another area of star formation that has been well studied by professional astronomers is the Triffid Nebula (M20), shown to the right. Professional images reveal this to be a stellar nursery, with 30 embryonic stars and 120 new-born stars discovered by NASA's Spitzer Space Telescope in 2005. One particularly notable feature that has been imaged very clearly by the Hubble Space Telescope is a gaseous jet around 0.75 light years long, being emitted by a hot young star. This is an example of a Evaporating Gaseous Globule, aptly dubbed an EGG, which will slowly be eroded by the strong radiation emitted by the young star.

Camera/scope: QHY23M with Skywatcher MN190 (1000mm f/5.2)
Exposures: H-alpha, Oxygen-iii, RGB (6x90seconds per filter all bin4)

A brilliant example of the impact hot young stars have on their progenitor nebula is the Orion Nebula. At the centre of the image is the Trapezium, the four principal stars that illuminate much of the surrounding nebula.

"We are all in the gutter, but some of us are looking at the stars"
Oscar Wilde

The Trapezium asterism, part of an Open Cluster at the heart of the Orion Nebula

Camera/scope: QHY163M with Skywatcher MN190 (1000mm f/5.2)
Exposures: LRGB (60x5 seconds Luminance; 55x10seconds per colour filter)

The Trapezium, officially designated ϴ1 Orionis, is part of an Open Cluster of stars, many of which are still hidden from view within the nebula. This is a really amazing group of stars, originally noted by Galileo as 3 stars in his observations from 1617, but in reality there are 5 principal stars, several of which are multiple star systems themselves!

ϴ1 Orionis A is actually a triple star system, the principal component being a B-class star
ϴ1 Orionis B has been proposed as a multiple system potentially containing 5 stars, the principal one a B-class star
ϴ1 Orionis C is a double star system, both being very large and hot O-class and B-class stars
ϴ1 Orionis D is large B-class star, which may have up to 3 companions

It fascinates me, to think that an image of only a few seconds duration through a modest amateur telescope showing four stars at the heart of the enormous Orion Nebula is actually a highly complex family of a dozen or more stars, within around 2 light-years of each other. Remember, our nearest neighbouring star is just over 4 light-years away, and of course our Sun has no companion.

The size of these stars is also something to contemplate and the impact they are having on the surrounding nebula. Classified by the Hertzsprung-Russell diagram (see page 37), these "four" stars are amongst the most massive types; O being the largest and B the second largest in terms of mass. It's worth noting that these are considered to be on the "main-sequence", or in the prime of life perhaps, with the stars being in hydrostatic equilibrium. In other words, the outwards radiative pressure generated by the fusion of hydrogen to helium is balanced by the inwards gravitational pressure of the star's mass. In a later chapter we will look at what happens to a star as it ages and moves away from the main sequence, becoming unstable.

High velocity stellar winds are emitted by these young stars, sculpting the surrounding nebula into the beautiful arches and curves evident in the long-exposure images we've already explored. The stellar wind is a flow of gas away from the star, driven by radiation pressure. Sometimes these high velocity winds from O and B-class stars sculpt beautiful bubbles in the surrounding gas, such as the aptly named Bubble Nebula (NGC7635) in Cassiopeia, which can be seen on the following pages.

Unfortunately, it's not possible to photograph the multiple star system components in the Trapezium using amateur imaging equipment due to their very close association. But we can do this with other famous systems, such as the Alcor-Mizar system in Ursa Major, which is actually a sextuplet star-system - albeit we can only image 3 of the stars with amateur equipment!

The observational history of Alcor and Mizar reveals much about the advancement of astronomy, with the earliest human records of Alcor and Mizar being used as a sight-test by the Romans around the 3rd century BC. Fast forward to 1617 and Mizar itself was discovered to be a binary system (Mizar A and B). Then in 1890 Mizar A was found to be a spectroscopic binary system, followed by Mizar B in 1908 and then as recently as 2009 Alcor was itself discovered to have a faint red-dwarf companion. Unknown to the Romans, Alcor and Mizar have also been confirmed as binary in 2009. So, a total of 6 stars all gravitationally related, 3 of which can be imaged by amateurs!

A wide-field view of the Bubble Nebula and Open Cluster M52 in the constellation of Cassiopeia

Camera/scope: Canon 300D with Takahashi E130D (430mm f/3.3)
Exposures: 18 x 300seconds @ ISO400

The wide-field picture above also includes data from the following close-up image to improve resolution of the nebula. This is a technique I use quite often now, to get both the broader context of the star field and the detail of the nebula. I love the colours of the yellow stars in this image, contrasting with the blues and whites of the hotter, younger stars.

"If the future does not include being out there among the stars and being a multi-planet species, I find that incredibly depressing"
Elon Musk

A close-up view of the Bubble Nebula (NGC7635) showing the expanding shell sculpted by the hot young star embedded in the gas cloud (upper left star within the bubble)

Camera/scope: QHY23M with Skywatcher MN190 (1000mm f/5.2)
Exposures: Ha-Oiii-RGB (12x300seconds Ha and Oiii each; 6x180seconds per colour filter)

The star that has produced this beautiful cosmic bubble is designated BD+60°2522 and is a bright O-class star. Its "surface" temperature is estimated to be around 37,000 Kelvin, which is pretty hot compared to our Sun's surface temperature of 5,700 Kelvin.

In astronomical terms the star can be described as young, at only 2 million years of age, or less than 0.05% the age of our Sun. Because it's so massive and hot, the strong stellar wind carries away huge amounts of the star's mass at around 2000km/s, perhaps as much as a thousand billion billion tonnes per year. At this copious rate of mass loss, the star is expected to live only a few million years before it ends its life spectacularly as a supernova. But more of that later.

The age of the star appears to be at odds with the estimated age of the Bubble Nebula itself, perhaps by a factor of a hundred, leading to one theory that the bubble is the result of a supersonic shock front interacting with an interstellar region of ionised hydrogen. It is generally thought that H-II regions, these areas of ionised hydrogen, are the second stage of development of a Giant Molecular Cloud wherein the atomic hydrogen (H_2) has been ionised by the strong ultraviolet radiation from the hot young stars born within the GMC. Around 90% of the H-II region is composed of this relatively hot ionised hydrogen, although they can be complex clouds that include a remainder of the nascent cold atomic hydrogen, as well as some parts that contain very hot plasmas in the region of the gas shock fronts associated with the strong stellar winds described above.

All of this goes to provide a complex photographic picture in many H-II regions with most images being dominated by the signature hydrogen-alpha emission line at 656.3nm, giving the characteristic red glow. But given the complexity of the cloud, there are often other emission lines readily accessible to the amateur photographer such as doubly-ionised oxygen (O-iii), ionised sulfur (S-ii) and of course the broadband spectrum too. Professional observatories, particularly the space-based telescopes, also image these regions for x-ray emissions, ultra-violet and infra-red radiation.

There are ultimately two consequences of these strong stellar winds sculpting the surrounding nebula. In some instances, the compression of the gas, caused by the shock front coming from the hot young stars, can trigger a new phase of star birth in other parts of the nebula where the conditions are right (i.e. the cooler neutral atomic hydrogen of the GMC). However, this doesn't happen in all cases. The other effect of the stellar wind and intense ultraviolet radiation is to disperse the nebula entirely over a period of 50-100 million years.

Another example of intense star formation within a complex GMC/H-II area is the region sometimes called the Cygnus Wall, which is part of the North America Nebula (NGC7000). The following image is a classic red-dominated hydrogen-alpha emission, but complex areas of dust in the foreground create the well-known shapes by blocking our line of sight to the main nebula.

Infrared images taken by the Spitzer Space Telescope have looked deep inside the dusty areas of the nebula and found clusters of young stars only 1 million years old. At this age, the young stars are likely to still be fusing deuterium in the outer shells of the core, but this fuel is limited and will be all used up within only a few million years.

"Look up at the stars and not down at your feet. Try to make sense of what you see, and wonder about what makes the universe exist"
Professor Stephen Hawking

The Cygnus Wall within NGC7000 (North America Nebula)

Camera/scope: QHY163M with Skywatcher MN190 (1000mm f/5.2)
Exposures: Halpha-RGB (6x300seconds Ha; 6x120seconds per colour filter all bin2)

The image combines broadband red, green and blue filters with a narrowband hydrogen-alpha filter to increase the contrast within the nebula. The h-alpha data is also mapped into the red channel of the image to increase the visibility of the nebula.

Over the page, a wide-field view of the whole of the North America Nebula shows how large this area is, at around 2 degrees x 2 degrees - remember the full moon is only half a degree in apparent diameter!

"Yeah we all shine on, like the moon, and the stars, and the sun"
John Lennon

Camera 1: QHY23M
Scope 1: TS INED70 0.8x reducer
Focal length: (336mm f/4.8)
Exposures: S-ii, H-a, O-iii
4x300seconds per filter (bin4)

Camera 2: QHY163M
Scope 2: Skywatcher MN190
Focal length: (1000mm f/5.2)
Exposures: H-a, RGB
(as detailed in previously for
the Cygnus Wall image)

Camera 3: Canon 300D
Scope 3: Takahashi E130D
Focal length: (430mm f/3.3)
Exposures: 13x300seconds
ISO800

This image is a combination of
three systems running at the
same time on my Skywatcher
EQ6 Pro mount. This setup is
nominally overloaded for the
mount, but with careful balancing
I have managed to ensure the
mount runs smoothly and
consistently achieves excellent
tracking (with auto-guiding)
over periods of 10 to 15 minutes.

I like the three-dimensional
aspect that is given by combining
multiple data sets with subtle
blending of the narrowband
and broadband data, presenting
a less intensely "red" image and
nicely coloured stars.

The California Nebula (NGC1499) is a wide H-II region in the constellation of Perseus extending over 2.5 degrees, emitting light in the h-alpha and h-beta wavelengths as the hot O-type star Xi Persei ionises the hydrogen. Taken on 4 nights in 2014, 2016 & 2017.

Camera/scope: QHY23M with Takahashi E130D (430mm f/3.3) & QHY163M with TS INED70 (336mm f/4.8)
Exposures: Ha-Sii-LRGB (9x600s h-alpha and sulfur-ii; 11x300s Luminance; 12x60seconds per colour filter)

Another example of a complex nebula is the large area of gas and dust in Cepheus that is catalogued as IC1396, which includes the Elephant's Trunk nebula, a dark globule containing a number of hot young stars.

HD206267

A two-panel mosaic of IC1396 and the Elephant's Trunk Nebula (right) and the bright, massive triple star system HD206267, responsible for ionising the nebula by harsh ultraviolet radiation

Camera/scope: QHY23M with Takahashi E130D (430mm f/3.3)
Exposures: Ha-RGB (5x300s h-alpha; 5x180seconds per colour filter)
Insets: QHY163M and Skywatcher MN190 (1000mm f/5.2); LRGB exposures (12x300seconds; 12x180seconds)

4 GROWING UP

In this chapter we will look at the third stage of stellar evolution, as stars come into their own, untying the apron strings as it were. The first image to tell this story is the Rosette Nebula, showing the progressive erosion of the home nebula from which the stars were born. Again, this is a combination of broadband and narrowband data.

"Keep your eyes on the stars, and your feet on the ground"
Theodore Roosevelt

Camera/scope: QHY163M with Takahashi E130D (430mm f/3.3)
Exposures: RGB (20x120seconds per filter)
Camera/scope: QHY23M with TS INED70 and 0.8x reducer (336mm f/4.8)
Exposures: S-ii, H-a, O-iii (5x600seconds per filter)

In this particular instance, the nebula is spherical in shape, seen by us in two dimensions as circular. And embedded at the centre is the cluster of stars with the label NGC2244. This object is around 5,000 light years away from us and measures 130 light years in diameter. There are around 2500 stars within the cluster, with the obvious bright stars at the centre being hot O-type stars that are shaping the nebula and dispersing it, particularly evident around the central stars.

There are numerous dark areas visible in the image, dusty cocoons that are still home to newly emerging stars. The narrowband data is mapped to the Hubble palette, with sulfur, hydrogen and oxygen mapped to red, green and blue respectively to give the classic gold and blue appearance. Although this is entirely unnatural in appearance, it does give a real depth to the image and allows the different elements to be displayed more prominently. Given that the image is not a true-colour view, it can be fun to play with it to get alternative renderings.

An alternative rendering of the narrowband data for the Rosette Nebula

But the best example in the northern sky of a cluster of stars that has completely outgrown its progenitor nebula is the Pleiades (M45), one of the most spectacular objects in the night sky.

" *Men are like the stars; some generate their own light while others reflect the brilliance they receive* "
Jose Marti

Camera/scope: Canon EOS 300D with Takahashi E130D (430mm f/3.3)
Exposures: 13x300seconds @ ISO400

There has been debate about whether the blue reflection nebula surrounding the Pleiades is the remnants of the birth cloud, or whether this happens to be a chance encounter between the young cluster and some unrelated interstellar gas. More recent theories favour the latter scenario. The cluster is clearly visible to the naked eye, with several bright components all of which are hot B-type stars. The most often used name for the cluster comes from the Greek legend of the seven sisters born to Atlas and Pleione, but there are legends from many cultures around the world linked to this bright and obvious cluster of stars.

The stars themselves are estimated to be around 100 million years old and there are over a thousand stars in the cluster overall, with 14 of them being naked-eye hot blue stars. The proper motion of the stars across the sky, along with statistical estimates of the likelihood of so many large young stars being so closely grouped in our sky, led to the conclusion in the 18th century that these were a cluster of genuinely associated stars, rather than just a line-of-sight grouping. The group is moving slowly across our sky towards the constellation Orion, and will eventually drift away from one another and disperse into space over a period of about 250 million years.

The principal reason that the nebulosity is no longer generally thought to be left over from the star-birth region is linked to the age of the stars (between 75-150 million years). By this age the hot young stars would have dispersed virtually all of the dust and gas from their progenitor nebula, leading to the conclusion that this is not the progenitor gas.

But why does this nebula shine with its strikingly blue colour, compared to many of the nebulae we've looked at with a strong red emission? Well, this is a classic example of a reflection nebula, not an emission nebula. In this instance the light from the bright stars is reflected by dust in the interstellar medium, the dust scattering the light. The colour of the reflection nebula is typically blue due to scattering being more efficient at the blue wavelengths than the red, although a proportion of the colour may also be linked to the predominant colour of the stars as well.

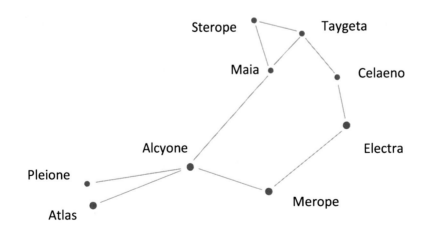

The Pleiades, seven daughters born to the mythical Greek Titan Atlas and the sea-nymph Pleione

"The Milky Way is nothing else but a mass of innumerable stars planted together in clusters"
Galileo Galilei

VdB152 (Cederblad 201) is a small faint blue reflection nebula embedded at the tip of a dark dust cloud

Camera/scope: QHY23M with Takahashi E130D (430mm f/3.3)
Exposures: LRGB image (8x600seconds Luminance; 11x120seconds per colour filter bin2)

VdB152 is another example of a wandering star lighting up a blue reflection nebula as it passes through unassociated interstellar dust. How do we know? The velocity of the bright blue star at the tip of the nebula is very different to that measured for the cloud itself, so this appears to be another chance interstellar encounter rather than a star and its associated birth nebula. Notably, in this nebula there is a subtle reddish glow from the dark dust cloud too, which is postulated may arise from ultraviolet excitation of the dust by the nearby star.

"Moonlight drowns out all but the brightest stars"
JRR Tolkien

The Iris Nebula in the constellation Cepheus is a lovely reflection nebula, surrounded by dark areas of dense dust

Camera/scope: QHY23M with TS INED70 and 0.8x reducer (336mm f/4.8)
Exposures: LRGB image (14x600seconds Luminance; 10x300seconds per colour filter)

The Iris Nebula (LBN487) and the small cluster of stars NGC7023 are a lovely sight, surrounded by dark interstellar dust clouds. The blue petals of the Iris itself extend over about 6 light years. The dusty nebula scatters light from the hot young star HD200775 (in the centre of the nebula) causing the blue glow.

36

It's worth remembering that the original size of the star dictates how long it will live, the biggest stars using their fuel at a prodigious rate and only living tens of millions of years, whilst the mid-sized stars live for billions of years and the smallest stars, such as red dwarfs, living for perhaps trillions of years on a very slow-burn. One of the best ways of visualising the lifecycle of a star and understanding the relationship between mass, temperature and relative size is the Hertzsprung-Russell diagram. Stars begin their life emerging as a stable object on the "main sequence", with low mass and lower temperature stars towards the bottom right of the curve and heavier, hotter stars towards the top left of the curve.

The relative luminosity, compared to the mass of our Sun is shown on the vertical axis. The temperature of the stars are shown ranging from hot blue through to cool red. Our Sun is a type of dwarf star with a spectral classification of G and a golden yellow colour.

As stars begin to age significantly, remembering that this is a relative term and can range from millions to billions or even trillions of years, they move away from the main sequence as they become unstable. How these stars develop depends on their initial mass. The really heavyweight contenders move off the main sequence towards the supergiant and hypergiant phase. Whilst the more moderate to small stars move off towards the subgiant and giant phases. And finally, the very low mass stars can simply run out of fuel and slowly cool. At the very bottom right corner of the diagram reside the brown dwarfs, classed as substars as they are not large enough to trigger hydrogen fusion, never going beyond the deuterium and lithium fusion stage.

Before we consider the later stages of a star's evolution though, let's take a look at some of the social groups that develop between stars. We've already seen how stars often have siblings, born from the same gas cloud. And we've looked at the Pleiades, a loosely associated cluster, where the stars have now drifted quite far apart.

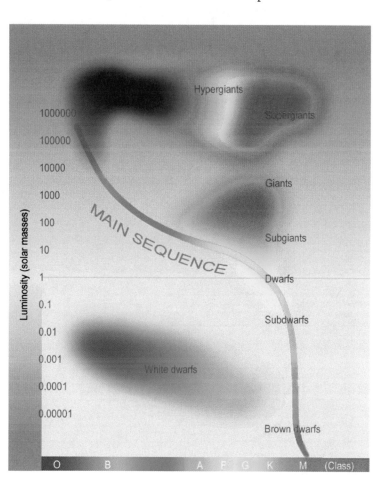

The Hertzsprung-Russell diagram

But what other kinds of groupings exist and how do these develop? Let's take a look next at Open Clusters, Globular Clusters and galaxies.

"We are all of us stars, and we all deserve to twinkle"
Marilyn Monroe

Messier 38 is an Open Cluster (left) and NGC1907 a small cluster (right)

Camera/scope: QHY23M with Takahashi E130D (430mm f/3.3)
Exposures: RGB image (5x300seconds per colour filter)

There are some interesting features in this image when you take the time to examine it in detail. I've added an annotated version of the image below to help.

First of all, the most prominent cluster is M38, a cluster that has passed beyond the flush of youth, at 290 million years of age. It contains about 100 stars and includes a bright yellow giant several hundred times the size of our Sun, denoted TYC2415-008-1. To the right of M38 in this image lies a much smaller cluster, NGC1907, which is approximately 500 million years old and contains only 30 stars. Both clusters are about 4500 light years away from us.

Just above M38 is a group of stars that are virtually undetectable as a cluster in this image, and is a much older grouping of stars catalogued as Czernik 21. They are estimated to be over a billion years of age and you can see that they have dissipated much more so than the two younger clusters.

Other objects of interest include a faint red nebulosity at the top of the image, catalogued as LBN794, and a very old "carbon star" catalogued as V* OU Aurigae appearing as a bright red dot in the image (also shown more clearly in the inset). The carbon star is on the asymptotic giant branch of the HR-diagram, having moved from being a main sequence star as it comes towards the end of its life. It has expanded in volume and cooled to become very red, as the image shows.

"Of all things visible, the highest is the heaven of the fixed stars"
Nicolaus Copernicus

Two very different star clusters feature in this image of M35 and NGC2158

Camera/scope: Canon 300D with Skywatcher MN190 (1000mm f/5.2)
Exposures: 14x120seconds @ ISO400

In this image, two very different clusters are visible. M35 is the younger one, at around 150 million years of age and containing lots of blue-white stars. Whilst NCG2158 is much older, with more yellow stars at around 2 billion years of age. These two clusters are not gravitationally associated, simply in a line of sight from Earth.

Going back to the metaphorical idea of stellar social groupings, Open Clusters could perhaps be thought of as a star's extended family of brothers, sisters, cousins, aunts, uncles, etc. They are all born from the same Giant Molecular Cloud, rather than a small cluster like the Pleiades which was probably formed in a smaller molecular cloud.

Another level of social interaction comes in the form of globular clusters, which could be described as distant relatives, as these tend to be some of the oldest stars found in a galaxy.

"How lovely are the portals of the night, when stars come out to watch the day die"
Thomas Cole

M13 Globular Cluster, in the constellation Hercules is one of the most photogenic objects in the northern sky

Camera/scope: QHY23M with Skywatcher MN190 (1000mm f/5.2)
Exposures: RGB image (8x300seconds per colour filter)

The M13 globular cluster was discovered in 1714 by Edmond Halley, although it was undoubtedly visible to humans from the earliest days of civilisation as it is of naked eye brightness from a dark sky location, at magnitude 5.8. Of course, before the advent of optics it would only have appeared as a faint fuzzy star. But a modest telescope or pair of binoculars reveals the structure of the cluster quite readily. And astrophotography really brings the cluster to life, although care has to be taken not to over-expose the core of very bright stars whilst trying to grasp all of the fainter members of the group.

The cluster contains several hundred thousand stars, which is a very significant increase on the typical number of stars in an Open Cluster. This globular cluster is around 145 light years in diameter and lies about 22,000 light years away from our Sun. The density of stars in this region of space is so significant that gravity leads to collisions between the stars of the cluster, forming new stars that are even hotter and bluer than the rest of the typical population in the cluster. These are nominally called blue stragglers as they appear at the very top left edge of the Hertzsprung-Russell main sequence line and are not considered to be part of the normal main sequence. There are alternative theories of how these types of blue stragglers form, including mass-transfer between two stars in a binary pair.

What is unusual about globular clusters is that they tend to reside in a halo around the centre of the galaxy, with the total population in our galaxy estimated to be around 180. But how these giant star-cities form is not very well understood still, even in the 21st century. They are typically composed of old low-metal content stars, indicating that these stars were born at the very earliest stages of the galaxy.

Some theories suggest that globular clusters may be the remnant cores of dwarf galaxies that have been cannibalised by the main galaxy, particularly for those super-massive globular clusters that contain millions of solar masses and, less typically, multiple stellar populations. The more typical globular cluster contains one type or population of stars, usually old and of low-metal content. However, some have bi-modal populations where newer star-birth has been triggered, perhaps by the merger of some of the original population. Or, as in the case of Omega-Centauri, the largest globular cluster in our own galaxy with over 10 million solar-masses, there are multiple populations within the cluster - perhaps indicating that this is a dwarf galaxy remnant where star-birth has occurred over an extended period in the more typical galactic manner leading to a range of star ages and types.

Interestingly, we can also image globular clusters in other galaxies, millions of light years away from us. One of the best targets to do this with is our largest galactic sibling, the Andromeda Galaxy (M31), located around 2.5 million light years away. Of course, at such a distance the individual globular clusters, containing hundreds of thousands of stars each, only appear as single stellar points to us in amateur images. Nonetheless, what a thought! We can image these stellar cities so far away from us with a simple backyard telescope.

The following image is a relatively close-up shot of part of the Andromeda Galaxy and I have marked the location of several globular clusters as identified by professional observatories. Also visible in the image are two much smaller satellite galaxies, M110 and M32, although in reality there are at least 14 satellite galaxies orbiting Andromeda.

The Andromeda Galaxy (M31) featuring numerous globular clusters

Camera/scope: Canon 300D with Skywatcher MN190 (1000mm f/5.2)
Exposures: 10x300seconds @ ISO800

When we consider the size of a typical globular cluster, containing perhaps 100,000 stars within a 100 light year region of space, the immensity of the galaxy suddenly comes into real perspective. These clusters, showing as single dots of light in the galaxy, are enormous volume of space at over 580 trillion miles across. How vast then is the galaxy itself! This particular galaxy is 220,000 light years across and contains over one trillion stars. Truly mind blowing.

The image above only shows some of the galaxy, so that we can resolve the globular clusters as points of light. The following image shows the full extent of this gargantuan continent of stars floating in intergalactic space.

"O, thou art fairer than the evening air clad in the beauty of a thousand stars"
Christopher Marlowe

The Andromeda Galaxy (M31) in its full glory, with satellite galaxies M32 and M110

Camera/scope: Canon 300D with TS INED70 and 0.8x reducer (336mm f/4.8)
Exposures: HDR image comprising (21x300seconds @ ISO400); (10x60seconds @ ISO400)
With additional h-alpha data to highlight areas of star formation in the galaxy (13x600seconds h-alpha)

I added some hydrogen-alpha data taken with the QHY23 camera and Takahashi E130D to give a little more definition to the areas of star-formation in the arms of the galaxy. These appear as brighter red areas of nebulosity in the image. Due to the brightness range of the galaxy, I also took a set of shorter exposures to enable the core to be processed without significantly over-exposing it.

Galaxies are the largest associated groupings of stars that we will explore in this book, with a few images of galaxies that are themselves gravitationally bound. The word galaxy comes from the Greek *galaxias* meaning "milky one" and is not dissimilar to the early scientific descriptions of galaxies as spiral nebulae. Galaxies come in all shapes and sizes, from the grand spirals like the Andromeda Galaxy, barred spirals like our own Milky Way, elliptical galaxies like M84, M86 and M87 in Markarian's Chain and a whole host of irregular shapes.

Many of the larger galaxies have attracted satellite dwarf galaxies, such as M32 and M110 orbiting the main Andromeda Galaxy M31. Our own galaxy has two very visible dwarf satellites, the Large Magellanic Cloud and the Small Magellanic Cloud, neither of which is visible from the north of England unfortunately, as well as another dozen less obvious ones!

How galaxies form is a controversial topic, with various theories that don't fully explain observations. One theory is that the galaxy formed from the singular collapse of a gigantic gas cloud, creating angular momentum as it contracted and cooled. An alternative theory is that multiple large clouds of gas clumped together and merged to form the galaxy. Current theories also speculate that dark matter may have a role to play in stopping the total collapse of the galaxy, as it remains in a halo around the outside of the galaxy. One thing that is certain, is that galaxies can change over time as they collide and merge with other galaxies, as can be seen in images like the Whirlpool Galaxy (M51) on page 52.

Furthermore, many of the large galaxies are also gravitationally bound, orbiting one another in a long slow dance lasting millions or billions of years. Our own galaxy is gravitationally bound to the Andromeda Galaxy, which is currently 2.5 million light years away but heading closer, and the Triangulum Galaxy (M33), which is about 2.7 million light years away. There is recent evidence that M31 and M33 have already had a strong tidal interaction sometime in the past 8 billion years, with streams of hydrogen gas observed trailing between the two galaxies. Current forecasts of motion predict another interaction within the next 3 billion years.

Similarly, observations of the relative velocity (speed and direction) indicate that the Andromeda Galaxy is likely to interact with our own galaxy in around 4 billion years. Either of these "collisions" would significantly alter the shape of the galaxies within the group, perhaps forming one or more giant elliptical galaxies in the distant future.

There are some interesting features in M33, including a very large H-II region similar in size to the Orion Nebula in our galaxy. The nebula was discovered by William Herschel and given a distinct reference number, which we know today as NGC604. Isn't it amazing that we can see areas of star formation in another galaxy almost 3 million light years away from Earth! This one is about 1500 light years across. In the next image, I have taken additional hydrogen-alpha data to supplement the colour image of M33 in order to show how many areas of active star formation exist in the galaxy. They appear as red/pink patches in the spiral arms of the galaxy. The brightest red patch at the left-hand side of the galaxy is NGC604. The galaxy has a low surface brightness and is more difficult to image than its sibling M31.

"If the stars should appear but one night every thousand years how man would marvel and stare"
Ralph Waldo Emerson

The Triangulum Galaxy (M33), one of the three major galaxies in our Local Group

Camera/scope: Canon 300D with Takahashi E130D (430mm f/3.3)
Exposures: 30x300seconds @ ISO400
Camera/scope: QHY23 with Takahashi E130D (430mm f/3.3)
Exposures: Ha+L for Luminance (6x300seconds h-alpha; 24x300seconds L) + HaRGB (8x180seconds per filter)

The next image is another classic grouping of galaxies, visible in the constellation Ursa Major. M81 is sometimes referred to as Bode's Galaxy and is the larger of the two main galaxies visible in this image. Its companion M82 is an unusual galaxy of significant interest. Lots of other galaxies are visible in this annotated image.

"It is the stars, the stars above us, govern our conditions"
William Shakespeare

M81 and M82 are a stunning pair of galaxies about 12 million light years away

Camera/scope: QHY163M with TS INED70 with 0.8x reducer (336mm f/4.8)
Exposures: LRGB image (19x60seconds Luminance; 20x60seconds per colour filter) + 9x600seconds h-alpha

M82 is a beautiful and unusual galaxy, with incredible bi-polar jets streaming out of the galaxy's core. This deep image shows the jets quite well, but they are extremely faint. Note the total imaging time is 7.25 hours and the colour data was binned 2x2, increasing the sensitivity of the camera by a factor of 3 to 4, the effect not being linear.

So what is going on in this galaxy that makes it so interesting, as well as visually stunning?

M82 is referred to as a starburst galaxy, due to the high rates of star formation, particularly in the dense core of the galaxy. In addition to the densely populated core, wherein there are many young massive clusters, M82 also sports a supermassive black hole at its centre. The rate of star formation is also linked to the rate of star death, or supernovae, almost certainly triggered by tidal interactions with its massive neighbour M81 around 200 million years ago.

On average, stars are being born and are dying 10x faster in M82 than in our own galaxy.

"If people sat outside and looked at the stars each night, I bet they'd live differently"
Bill Watterson

M82 is a highly active "starburst" galaxy with enormous jets of gas flowing away from its core

Camera/scope: QHY23M with Skywatcher MN190 (1000mm f/5.2)
Exposures: Ha-RGB image (21x900seconds H-alpha; 4x600seconds per colour filter bin2)

"The sun, the moon and the stars would have disappeared long ago... had they happened to be within the reach of predatory human hands"
Havelock Ellis

M81 is a beautiful spiral galaxy, the largest in a cluster of 34 galaxies

Camera/scope: QHY23M with Skywatcher MN190 (1000mm f/5.2)
Exposures: LRGB image (10x180seconds Luminance; 10x180seconds per colour filter bin2)

M81 is sometimes referred to as a grand design spiral galaxy, having well defined spiral arms that extend around the galaxy in an unbroken swirl, in this case going right from the centre out to the far edges. The formation of these beautiful structures is thought to be linked to close encounters with companion galaxies, such as M82 and NGC3077.

In the previous image, visible to the top left of the galaxy is a small, faint patch of stars, catalogued as Holmberg IX, which is actually a small irregular satellite galaxy orbiting M81. This is thought to be around 200 million years old, making it very young, galactically speaking! The inner parts of the main galaxy show a predominance of older yellow and red stars, with the outer regions tending to be populated with hot blue younger stars. At the centre of the galaxy resides a supermassive black hole of around 70 million solar masses, approximately 15 times the size of the one at the centre of our own Milky Way galaxy. The much younger and fainter satellite galaxy, Holmberg IX was only discovered in 1959, whereas M81 was discovered nearly 200 years earlier in 1774 by J E Bode.

As a teaser for the final chapter on the stellar lifecycle, death and rebirth, I've included the following widefield view of M81 and M82 featuring a supernova that was observed in M82 on 21st January 2014. It could be said that all stages of the stellar lifecycle are visible in this one image - areas of ionised gas in the arms of M81, young stars in Holmberg IX, older stars towards the centre of M81 and the death throes of a single star going supernova in M82, apparently emitting as much light as the entire galaxy for a brief time. But more of that in Chapter 6.

The interesting back story to the discovery of this supernova, dubbed SN2014J, is that it was officially discovered by a team of students and their tutor from University College London using a relatively small 14" diameter telescope.

The supernova appears as a bright yellow star to the left of the core in this image. Recall that this galaxy is 12 million light years away and this explosion is in that galaxy, truly immense power! All the other stars visible in this image are in our galaxy.

Compare this image to the one on page 48, to see that the supernova only appears in the later 2014 image.

Another stunning spiral galaxy found in the constellation Ursa Major is the Pinwheel Galaxy, catalogued by Charles Messier as M101. This face-on spiral is reminiscent of a spinning pinwheel firework, and indeed it has lots of H-II areas of star formation blazing away in its arms, visible in this image as glowing red/pink areas.

"Who among us has never looked up into the heavens on a starlit night, lost in wonder at the vastness of space and the beauty of the stars?"
Jeb Bush

The Pinwheel Galaxy (M101) is 21 million light years away, over 170,000 light years in diameter and contains 1 trillion stars

Camera/scope: QHY163M with Skywatcher MN190 (1000mm f/5.2)
Exposures: Ha-L HaRGB image (10x900seconds h-alpha; 8x120seconds Luminance and per colour filter bin2)

The nebulae are brought to life with hydrogen-alpha data added in to the Luminance and red colour channels.

The next galaxy in our collection is the famous Whirlpool Galaxy, or Messier 51. This face on spiral has a companion galaxy, NGC5195 which it is interacting with very clearly. The two galaxies are connected by a dark dense dust lane and the tidal interaction between the two is very clear with the sweeping tails of stars surrounding them.

Camera/scope: QHY23M with Skywatcher MN190 and 2.5x converter (2500mm f/13)
Exposures: LRGB image (12x300seconds Luminance; 12x180seconds per colour filter all bin2)

This object is actually located in the constellation Canes Venatici, but is just below the "left hand" star of Ursa Major, making it pretty easy to find.

Again, the grand spiral structure of M51 is thought to be influenced by the strong interaction with its companion, also leading to relatively high rates of star formation. One theory is that the smaller galaxy passed through the middle of the larger one around 500 million years ago.

Recent observations by professional astronomers have revealed unusual arcs of energetic material emitting X-rays in the centre of NGC5195, possibly caused by gas within the galaxy being channelled into the supermassive black hole at the centre of the galaxy.

Whirlpool Galaxy (M51) and its companion NGC5195 in Ursa Major

This is one of my earliest images with the QHY23M CCD camera, and I was experimenting with the addition of a focal extender (2.5x in this instance) and the use of pixel binning to compensate for the reduced brightness of the image that occurs when the focal length is increased in this way. The image is a little soft and would benefit from significantly more data being gathered to give better colour fidelity and more detail in the fainter areas of the image.

M106 is a spiral galaxy 22-25 million light years away and 135,000 light years in diameter

"A philosopher once asked, 'Are we human because we gaze at the stars, or do we gaze at them because we are human?' Pointless, really...'Do the stars gaze back?' Now, that's a question!"

Neil Gaiman

Camera/scope: QHY23M with Skywatcher
MN190 (1000mm f/5.2)
Exposures: Ha-LRGB image (5x300seconds H-alpha;
12x300seconds Luminance and per colour filter bin2)

M106 shows a very unusual feature, just visible in this relatively short image. The galaxy has a pair of arms that appear to be inclined at an angle to the rest of the galaxy and instead of comprising a mix of stars and gas, these arms are purely hot gas, glowing red. Only one of the arms is visible in my image, extending up from the left-hand side of the galaxy core. The other arm extends down to the bottom from the right-hand side of the core. For a fantastic view of this feature, look up M106 on the NASA website, where you can see an image created using amateur images mixed with Hubble data. It's truly gorgeous!

This galaxy was discovered by Pierre Mechain in 1781 and is located about 24 million light years away. The supermassive black hole at the core of the galaxy is thought to be responsible for stirring up the gas in these odd arms, ionising it and expelling it out of the plane of the galaxy. This is one of the classic Seyfert galaxies, named after Carl Keenan Seyfert who identified this class of galaxy, a sub-class of those galaxies with active galactic nuclei.

Characteristics of a Seyfert galaxy are the presence of a supermassive black hole at the core, surrounded by an accretion disk which is feeding material into the black hole. These galaxies usually have traditional spiral arms, but also have a very compact and super-bright core, particularly when observed in the ultraviolet portion of the spectrum.

Another unusual feature of this particular galaxy, M106, is the presence of a *water megamaser*. What on earth is that, you might ask? A maser is similar to a laser, but rather than operating in the visible wavelengths, masers operate in the microwave region of the electromagnetic spectrum. So, maser stands for Microwave Amplification by Stimulated Emission of Radiation. But what is stimulated emission? This is when atoms or molecules absorb energy from a source of electromagnetic radiation and become 'excited', giving them a boost from a lower energy level to a higher energy level. This can lead to the emission of a photon at a specific wavelength, with a directional focused beam. There are specific conditions that must exist for a natural maser to occur, such as the density of the masing molecule, a suitable source of radiation energy and other more complex environmental factors. And so, the detection of a maser or megamaser can give insights into the nature of the interstellar medium, can help measure the distance to galaxies (independently of the supernova-based standard-candle method) and can help to determine the mass of the black hole at the centre of galaxies such as M106.

Drawing this chapter to a close, the last image zooms out beyond the idea of a single galaxy, or even a few galaxies tied in an intergalactic dance, to an even larger scale where dozens, hundreds or myriads of galaxies can be seen in far off space. We know from images like the Hubble Ultra Deep Field that the more we look, the more we see, with countless galaxies extending beyond our visual time-limited horizons. The following image is a far cry from such famous masterpieces, but it does demonstrate how vast our universe is when you consider that each galaxy contains millions or billions of individual stars. I've only labelled a few of the more prominent galaxies in the image, but I've counted over 30 smudges in this short image of only 20 minutes (Luminance).

Our next chapter will look at what happens as stars move from being middle-aged to old aged.

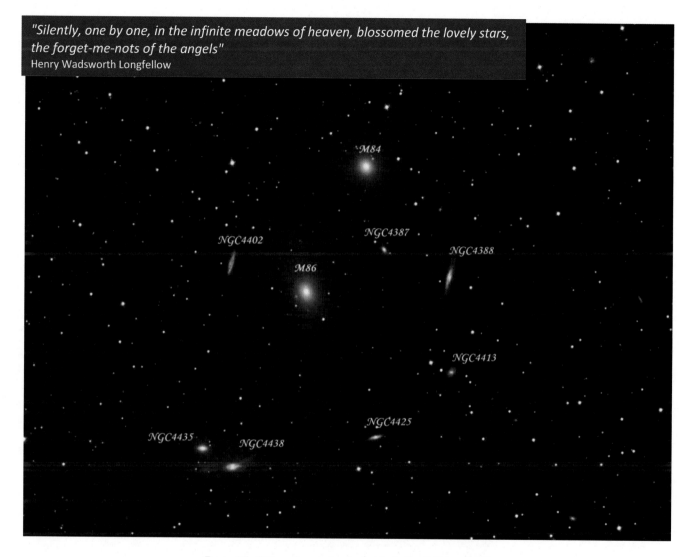

"Silently, one by one, in the infinite meadows of heaven, blossomed the lovely stars, the forget-me-nots of the angels"
Henry Wadsworth Longfellow

Part of the Virgo cluster of galaxies, including M84 and M86

Camera/scope: QHY23M with Takahashi E130D (430mm f/3.3)
Exposures: LRGB image (2x600seconds Luminance; 2x300seconds per colour filter bin2)

5 OLD AGE

What happens to a star's nuclear fusion process as it develops into middle age? Again, the answer depends on the original size of the star. But for the sake of simplicity, the following description can be taken to be sufficiently generic to illustrate what happens as a star ages. Recall that the earliest stages of nuclear fusion begin with deuterium, which is a stable form of hydrogen. The low temperature reaction (between 1 and 4 million Kelvin) is largely self-regulating due to convection within the star, allowing the star to accrete more matter from its birth nebula. This first stage is part of the proton-proton chain reaction that dominates in smaller stars. As the star continues to grow in mass, the pressure and temperature in the core will rise, and once it reaches between 10 and 17 million Kelvin hydrogen fusion can occur. In higher mass stars (>1.3x the mass of the Sun), the carbon-nitrogen-oxygen (CNO) fusion cycles dominates but some parts of the star will continue to follow the proton-proton reaction chain, although the hotter the core the more efficient the CNO cycle becomes.

Let's just remind ourselves of what's happening in this big ball of gas. At this stage, it's predominantly hydrogen gas. Imagine you were swimming about in this gas and decided to dive down deeper and deeper towards the centre of the ball. As you dive deeper, the pressure goes up - you can feel it in your ears! As the pressure increases, the temperature of the gas also increases in line with the ideal gas law, which says that for a given volume the temperature increases if the pressure increases. Once the gas reaches the critical temperature/pressure, both reaction methods effectively turn hydrogen into helium, albeit in different ways, releasing enormous amounts of energy in line with Einstein's famous equation $E=mc^2$.

In its inner "core" our Sun (which is considered a small star) converts around 600 million tonnes of hydrogen into helium every second, which sounds like a huge amount, but the mass of the Sun is about 1,988 million million million million tonnes and about 75% of it is hydrogen, so it's expected to last about 9-10 billion years!

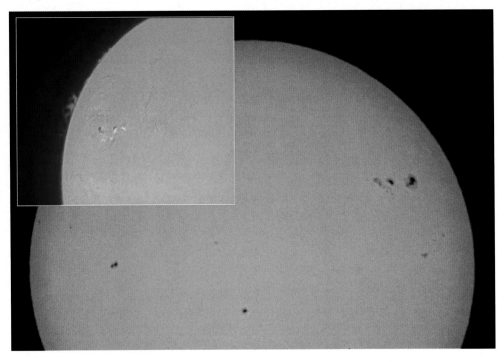

Our Sun, in visible light showing sunspots and in hydrogen-alpha showing prominences

The previous images are taken with special filters to reduce the intense light of the sun. The larger image uses a specialist high density mylar filter over the front of the telescope. The smaller inset image is taken with a dedicated solar telescope which uses an etalon filter to restrict the amount of light and the wavelength of light passing through the telescope. In this image, the scope is a Coronado PST 40mm telescope and it shows flares on the limb of the sun erupting into space, focusing on a very narrow 1Angstrom portion of the hydrogen-alpha wavelength.

The core temperature of our Sun is modelled to be 15.6 million Kelvin, so the dominant process of nucleosynthesis is the proton-proton chain, although as the Sun gets older it is expected that the CNO cycle will become more important. Note, only the inner core is hot enough to support hydrogen fusion, so the outer region is just a radiative and convective mass of gas. But as the fusion process ticks along over billions of years, the amount of helium in the inner core increases and eventually the star will begin to fuse helium into carbon and oxygen in its core, and hydrogen into helium in its outer shell. For stars like our Sun, or smaller, this is where the nucleosynthesis ends, as the helium fusion phase eventually ceases. Some interesting and beautiful nebulae are created during these later unstable stages of the small star's life, as it expands into a red-giant. A strong stellar wind blows away a large proportion of its outer layers in a steady stream of gas and then periodically the unstable star ejects large amounts of gas to form a planetary nebula. This can occur several times, leading to complex spherical shells of gas that interact with earlier ejections from the star, such as the Ring nebula.

The Ring Nebula is 2,300 light years away and the shells of gas are expanding at 20-30km/s. The central star is a faint white dwarf, which is the destiny of our Sun in the next 5 billion years once it runs out of hydrogen and helium. The white dwarf no longer sustains nuclear fusion, so the gravitational pressure is not counteracted by radiative pressure, causing the star to collapse to a dense core of carbon and oxygen.

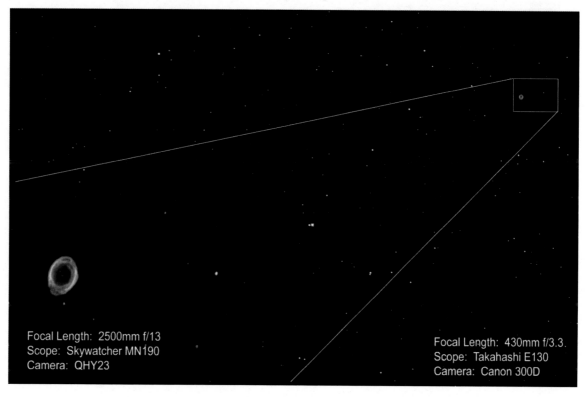

Focal Length: 2500mm f/13
Scope: Skywatcher MN190
Camera: QHY23

Focal Length: 430mm f/3.3.
Scope: Takahashi E130
Camera: Canon 300D

Another beautiful bipolar planetary nebula of the same kind, is the Dumbell Nebula (M27) in the constellation Vulpecula. This one is much larger in our sky than the Ring Nebula at 8 arc-minutes in diameter. Astronomers calculate that the nebula has been expanding for around 10,000 years. The physics of how these bi-polar nebulae are created is still uncertain, but their beauty is undeniable.

"It's the kind of kiss that inspires stars to climb into the sky and light up the world"
Tahereh Mafi

The Dumbell Nebula (M27) showing the bi-polar structure with multiple inner and outer shells of gas

Camera/scope: QHY23M with Takahashi E130D (430mm f/3.3) and Skywatcher MN190 (1000mm f/5.2)
Exposures: Two sets of data combined including LRGB and H-alpha/O-iii filters totalling 3 hours

Some other examples of planetary nebulae include Thor's Helmet (NGC2359) in Canis Major, low on the southern horizon from the north of England; and the Owl Nebula (M97) riding high in the northern sky - both pictured below.

The Owl Nebula (M97)

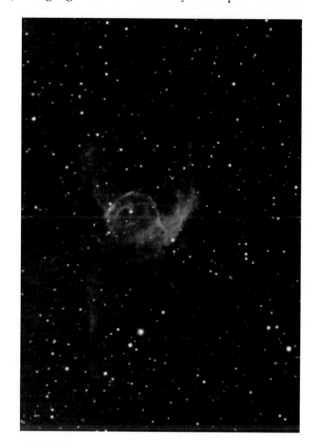

Thor's Helmet (NGC2359)

Camera/scope: QHY23M with Skywatcher MN190
Exposures: S-ii; O-iii; H-alpha image
(6x300seconds per filter bin2)

Camera/scope: QHY23M with Skywatcher MN190
Exposures: H-alpha; O-iii; RGB image
(5x600seconds Ha & Oiii; 2x60seconds per colour filter bin2)

The image of Thor's Helmet has very little colour data (RGB), so I added the much longer narrowband data into the colour channels as well as using them for a combined Luminance layer. The main benefit of doing this was to overcome some of the terrible atmospheric seeing conditions so low on the horizon. This object only rises to 20º altitude from my northern latitude in Cumbria. Also, I took this image when the moon was 80% illuminated, which is another major benefit of narrowband imaging compared to normal colour photography.

The Owl Nebula was discovered in 1781 but was only dubbed the Owl Nebula after it was observed and sketched by William Parsons, the 3rd Earl of Rosse in 1848. It's an obvious moniker given its appearance! The nebula is estimated to be 8,000 years old and lies about 2,000 light years away from us. This is an example of a star of similar size to our Sun, perhaps a little larger, that has evolved along the asymptotic giant branch, and perhaps underwent a very brief "helium-flash" ejecting hydrogen, helium, nitrogen, oxygen and sulfur into surrounding shells of material. As the star then progressed to becoming a stable white dwarf, with around 50-60% of the Sun's mass, its surface temperature of over 30,000 Kelvin was hot enough to emit ultraviolet radiation that ionised the previously ejected matter making it emit light itself.

The Owl Nebula has a radius of just under 1 light year and it is estimated to contain 0.13 solar masses of material from the progenitor star. This is relatively small in terms of mass for a planetary nebula, although it is typical of the physical size. So, the density of this particular nebula is very low at around 100 particles per cubic centimetre, compared to younger planetary nebula with a million particles per cubic centimetre.

By contrast, Thor's Helmet is a different kind of nebula, perhaps arguably not a planetary nebula at all. The structure is much larger at around 30 light years across and is probably a combination of material ejected from the progenitor star (in the same way a planetary nebula is formed) interacting with an unrelated molecular cloud. The central star is an example of a Wolf-Rayet star (WR-7), which is around 16 times the mass of our Sun and expected to progress its evolution to become a supernova. The intense hot winds from the star are sculpting and ionising the bubble shape in the gas.

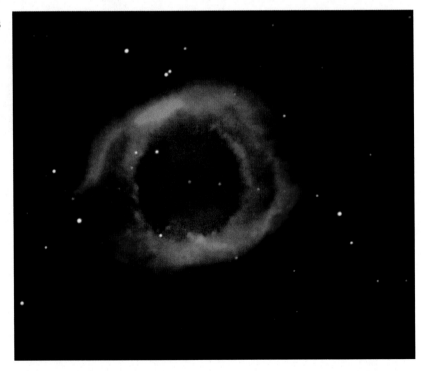

Another beautiful planetary nebula, which is only 700 light years away from Earth, is the Helix Nebula (NGC7293) also sometimes called the Eye of God for obvious reasons. This is another example of a dying low mass star that is transitioning to become a white dwarf, throwing layers of gas off in shells. The layers have become ionised by the exposed core of the star, causing them to emit light in h-alpha and oxygen-iii.

Interestingly, towards the centre of the nebula in the region where the red meets the blue there are a lot of dusty pockets of neutral molecular gas called cometary globules. They have elongated tails streaming away radially from the centre of the nebula, much like a comet. Over 20,000 of these have been discovered in the Helix nebula alone, each the size of our solar system!

To recap, we've looked at the evolution of low to medium mass stars, which are limited to two main stages of fusion, hydrogen to helium and helium to carbon/oxygen. But what happens to higher mass stars, such as the one that created Thor's Nebula?

Perhaps unsurprisingly, the nucleosynthesis progresses through the periodic table of elements, moving through progressively heavier elements. (It may be a glib statement, saying that this perhaps appears to be fairly obvious in our modern age, but I mean no disrespect to the eminent thinkers who first developed these theories. In fact, the idea of hydrogen-helium fusion was first proposed less than a hundred years ago by Arthur Eddington in 1920. The theory was built on through papers by Bethe and Weizsacker in 1938/39, followed by a more detailed description of the full fusion chain by Fred Hoyle in 1946 and 1954.)

The simplified perspective of the physics driving this process is the same as for the lower mass stars; the temperature and pressure in the core of the star continue to rise and as key trigger points are reached a new phase of nuclear burning commences in the heavier element. It's simplest to picture the star as an onion. At the start of its life, this higher-mass star (typically more than 8x the mass of the Sun) is largely hydrogen, with perhaps some helium and a small proportion of heavier elements depending on the make-up of the progenitor nebula.

Initially, hydrogen is fused into helium in the core of the star, but this time using the CNO process rather than the proton-proton chain since the core temperature is much hotter at over 17 million Kelvin. As this continues, a core of helium begins to build up, with the lighter hydrogen being pushed into the outer layers. Once the hydrogen-fusing core is exhausted, the helium core contracts and the temperature rises to 100 million Kelvin, triggering helium fusion to carbon with some oxygen produced.

Next in line, carbon fusion occurs at around 500 million Kelvin as the helium at the core is exhausted, creating oxygen, neon, sodium and magnesium.

Neon burning kicks in at 1.2 billion Kelvin generating more oxygen and magnesium. Again, as the central core of neon is exhausted, the core contracts further raising the temperature to around 2 billion Kelvin triggering oxygen fusion that produces silicon and sulfur predominantly, although by-products include chlorine, argon, potassium and calcium.

The final stage of nuclear fusion in this part of a star's life is silicon fusion, which occurs at around 3 billion Kelvin and creates iron as well as titanium, chromium, nickel and zinc. The reason this is the last stage is because after iron, the elements require more energy than is given out by any fusion, hence the chain reaction stops.

But what about the onion? All through this sequential process the previous reaction process is pushed into an outer layer where the fusion of that earlier element continues. So, the process builds up shells of elements that continue to undergo fusion in that layer, whilst the new core element begins its own fusion process.

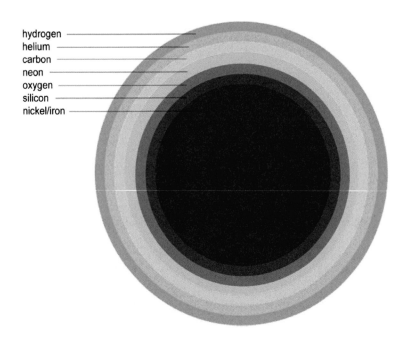

hydrogen
helium
carbon
neon
oxygen
silicon
nickel/iron

The time it takes for these different stages of fusion to progress depends on the initial size of the star, but the relative duration of each stage is roughly:

hydrogen - 10s of millions of years
helium - millions of years
carbon - 1000s of years
neon - 3 years
oxygen - 100 days
silicon - 1 day
nickel/iron - minutes

The diagram is just to illustrate the point and is not to scale.

The onion-like shells of matter created by stellar nucleosynthesis in large stars

So to finish this stage of the lifecycle with an image, the Crescent Nebula in Cygnus is another example of a complex emission nebula that has been created by a rare Wolf-Rayet star. The central star (WR-136) is approaching the end of its life, at 4.7 million years of age. It was a large star from birth and burned its fuel quickly, then expanded to become a red supergiant as it ran out of fuel and is now close to death, living out its last few hundred thousand years as a Wolf-Rayet star.

It is extremely hot with a surface temperature of around 70,000 Kelvin and fifteen times the mass of the Sun. The energetic stellar wind coming from the star is travelling at around 1,700km/s, colliding with the outer shells of gas ejected during the supergiant phase and sculpting a bubble shaped shock front.

Once it has progressed through the remaining stages of fusion, this star is expected to end its life as a spectacular supernova - in the "near" future. Given that the star is 5,000 light years away from Earth, perhaps it has already gone supernova and we just haven't seen it yet!

"The stars are like the trees in the forest, alive and breathing"
Haruki Murakami

The Crescent Nebula (NGC6888) in Cygnus, surrounded by wispy clouds of gas and dust

Camera/scope: QHY23M with Takahashi E130D (430mm f/3.3)
Exposures: H-alpha-RGB (3x600seconds H-a; 6x180seconds per colour filter bin2)

6 DEATH & REBIRTH

This chapter will look at what happens once the star has run out of fuel. As a visual reminder of the processes we've looked at so far and to introduce the final fate of stars, both small and large, I've put together this diagram using my own images and a little artistic license.

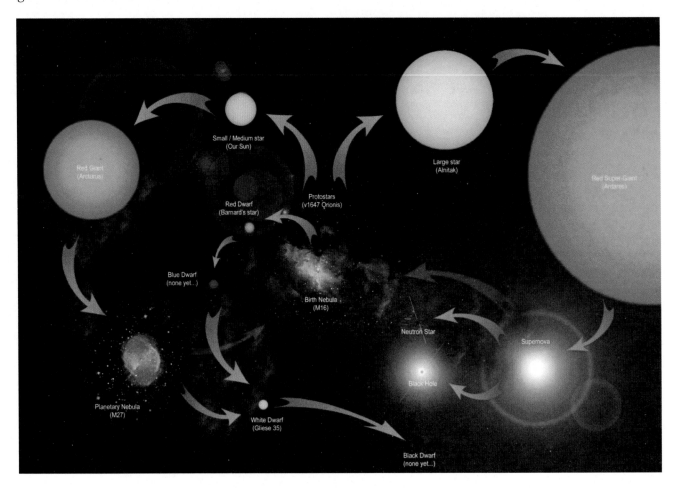

The full lifecycle of a star, from birth to death and re-birth

The diagram shows three routes for stars as they grow from the protostar, either as very small (red-dwarf route), small-medium (the Sun) or large stars (such as Alnitak in Orion's belt).

Following the smallest route first, a red-dwarf will burn its fuel very slowly and last for trillions of years, or so we theorise as the universe is not yet old enough to contain any "old" red-dwarfs by a long way, let alone any blue-dwarfs!

The medium sized stars develop into red giants, passing through the planetary nebula stage and then decay away to a white dwarf. Again, over very extended periods of time it is theorised that the white dwarf will finally cool to the point where it no longer emits radiation, referred to as a black dwarf. None of these exist yet, or will for a very, very long time.

The third route is for the large stars. Large really meaning anything over about 8 solar masses as a generalisation. Once the large stars begin to run out of fuel they expand to form a red super-giant and finally end their days in a cataclysmic supernova event. Depending on the size of the progenitor star, the supernova remnant will either be a neutron star or a black hole.

The evolution of the stars, particularly the later stages where layers of the star are ejected to form planetary nebulae and most especially the violent death of larger stars in supernovae lead to an enrichment of the galaxy in terms of heavy elements. Indeed, without some of the more spectacular events the heaviest elements simply wouldn't exist. Hence, the picture shows the creation of new nebulae by the mid-to-late stages of a star's lifecycle, enriched with heavier elements and restarting the whole process of star birth.

So what sort of images can amateurs take of these later stages of stellar evolution?

- Supernova events themselves
- Supernova remnant degenerate stars, such as the neutron star at the centre of the Crab Nebula (M1)
- Supernova remnant nebulae that have been enriched and often triggered into new phases of star birth by some of these spectacular stellar deaths

Images of supernova events are not visually spectacular, despite the immense power output of the process, mainly because they are typically so far away from us. There is one candidate supernova remnant that has been discovered near the Vela supernova remnant, which may have occurred around 900 years ago, although there is no record in our human history of this being observed. It's estimated that a supernova event, which could have an impact on our ecosystem due to the high radiation outputs, might occur once every few hundred million years. These events would need to be within around 30 light years, noting that some of the closest occurrences have only been 500 - 800 light years away, occurring tens to hundreds of thousands of years ago.

So, a typical image of a supernova is the brightening of an "invisible" spec of a star to a bright point of light in a galaxy far, far away.

Supernova SN2012aw (discovered March 16th 2012)
by Paulo Fagotti (Italy)

Galaxies M95 and M96 and the stunningly bright Mars. Supernova 2012aw in M95 discovered in 2012 by an amateur.

Camera/scope: Canon 300D with Skywatcher MN190 (1000mm f/5.2)
Exposures: 30x300seconds @ ISO800

There's a really interesting story behind this particular supernova. The galaxy M95 is only 37 million light years away, so it is close enough for our large professional observatories to image individual features within the galaxy. So, when this supernova event occurred in 2012, astronomers searched back through Hubble and other ground-based observatory archives to identify the likely progenitor star that we observed ending its life in March 2012. I took this image a few days after the supernova was discovered by Paulo Fagotti et al.

The image is unusual because it contains a super bright planet in the same field of view as the two galaxies. I was actually just interested in photographing the conjunction between Mars and the two galaxies, unaware that a supernova had been discovered only two days earlier by amateur astronomers. So, what do we know about the star that ended its life so spectacularly? A paper by Fraser et al (Red and Dead: The Progenitor of SN 2012AW in M95, *Astrophysical Journal Letters*) suggests that this was a red supergiant of between 14 and 26 solar masses based on archival images of the area taken by Hubble and other infrared ground-based systems. The mass of the star has been refined in a 2016 paper to between 11 and 14 solar masses.

But I'm getting slightly ahead of myself. What exactly is a supernova? How does it develop and what role does it play in the evolution of stars in general?

There are two basic types of supernova, one triggered by the core collapse of a massive star as it runs out of fuel (Type II supernova or Type Ib/c); and another where a degenerate white dwarf star manages to accrete matter from a nearby companion star, triggering a runaway nuclear explosion once it reaches the Chandrasekhar mass limit of 1.44 solar masses (Type Ia supernova). Both types of supernova produce incredibly powerful explosions that trigger the creation of elements heavier than iron, remembering that nickel/iron was the end of the normal nucleosynthesis process in large stars.

Recall that the nuclear fusion chain that develops in the massive stars gets faster and faster, creating heavier elements until it gets to the point where the chain suddenly stops because it requires more energy to fuse elements heavier than iron than is given out by the fusion process itself. There are several ways that the final core collapse can be triggered, but it's simplest to focus on the scenario where the iron core exceeds the Chandrasekhar mass limit, meaning that the gravitation pressure from the mass of the core overcomes the *electron degeneracy pressure* (which is a physical manifestation of the Pauli exclusion principle whereby two electrons cannot occupy the same quantum state at the same time).

The core itself collapses at phenomenal speeds of around 70,000km/s or 23% of the speed of light! This drives a dramatic increase in pressure and temperature in the core compressing the matter into a sea of neutrons at over 100 billion Kelvin. The neutrons also behave in accordance with the Pauli exclusion principle, exhibiting *neutron degeneracy pressure*, stalling the core collapse any further and causing a rebound shockwave outward from the neutron core coupled with a neutrino pulse of a few milliseconds, together driving the main supernova explosion. The hugely elevated temperatures trigger the formation of lots of heavy elements, which are then distributed into interstellar space by the violent explosion.

Following the supernova explosion, and depending on the mass of the original star, a neutron star will form in which all of the matter is a compressed ball of neutrons, typically only 20 kilometres in diameter with around two solar masses. These are very exotic objects, often generating huge angular momentum through their collapse, rotating at significant percentages of the speed of light. They are the densest stars known, they have the strongest magnetic fields and some of the highest temperatures and pressures. But in terms of mass, they are limited to around 3 solar masses, beyond which the progenitor star would actually break the neutron degeneracy pressure and lead to the creation of a black hole.

"The cosmos is within us. We are made of star-stuff. We are a way for the universe to know itself."
Carl Sagan

The Crab Nebula in Taurus is a supernova remnant, hosting a neutron star at its heart

Camera1/scope1: QHY23M with Skywatcher MN190 and 2.5x converter (2500mm f/13)
Camera2/scope2: QHY163M with Skywatcher MN190 (1000mm f/5.2)
Exposures (1): H-alpha; O-iii; RGB (5x600seconds H-a; 4x600seconds O-iii; 3x120seconds per colour filter bin4)
Exposures (2): H-alpha; O-iii bi-colour image (40x30seconds H-a; 20x30seconds all bin2)

This image of the Crab Nebula uses data from two separate camera and scope combinations with very different techniques. The first image combines long duration narrowband exposures (hydrogen-alpha and doubly-ionised oxygen) with standard red, green and blue data at a long focal length to capture detail in the nebula. The second image uses much shorter exposures with an ultra-low noise CMOS camera (the QHY163M) in high-gain mode and a higher number of frames but with only 30 second exposures. Note, that in both images I have boosted the sensitivity by binning the pixels, 4x4 in the 2500mm f/13 rig, and 2x2 in the 1000mm f/5.2 rig. By using the 2.5x teleconverter, more detail is resolved but at the expense of light gathering power, hence the need to bin the pixels 4x4, which theoretically increases the sensitivity by a factor of 16 in comparison to an unbinned exposure, although in practice it's less.

The Crab Nebula itself is one of the most famous supernova remnants in the sky. The Chinese recorded the eruption of a bright new star in the sky in the year 1054 AD, which was more latterly identified as a supernova and associated with this beautiful nebula around 1921, almost 200 years after the initial discovery of the nebula. The nebula is now 11 light years across and has been calculated as expanding at the rate of 1,500km/s!

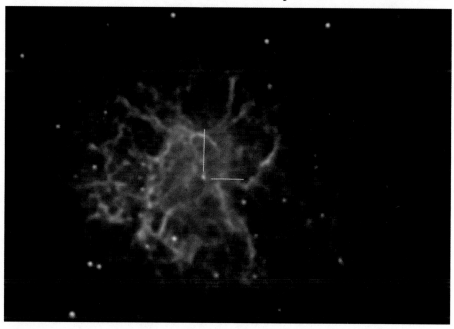

The progenitor star here is thought to have been between 8 and 12 solar masses, which collapsed as a Type II core collapse supernova. Modern observations have detected the presence of hydrogen, helium, carbon, oxygen, nitrogen, neon, sulfur and iron in the ionised nebula, very much in line with the theory of stellar nucleosynthesis in large stars as we have already discussed. Whilst at the centre of the nebula, an unusual star was discovered in 1942, as a strong source of radio and X-rays. A little later, in 1968, the star was identified as the first Pulsar. This was later understood to be a rapidly rotating neutron star, sending out focused beams of radiation that are seen from our perspective as regular, rapid pulses.

The space-based Chandra X-ray telescope has taken some fantastic images of the nebula, revealing a three-dimensional structure shaped a bit like a doughnut with powerful shock fronts and polar jets emitted by the pulsar. Repeat images of the inner portion of the nebula have shown it to be a very dynamic place, with the structure changing over very short timeframes of only a few months.

I have marked the location of the magnitude 16 pulsar neutron star in this close up image.

The next set of images will showcase the Veil Nebula, in the constellation Cygnus. This is another classic imaging target for amateur photographers, extending over a very wide area of sky at over 3 degrees apparent size. The nebula has spread out over about 70 light years since the supernova eruption around 8,000 years ago, with each main component having its own catalogue designation and familiar name.

The western portion (right) is the Witches Broom (NGC6960), the central region is Pickering's Triangle and NGC6979/6974, and the eastern portion (left) is sometimes called the Network Nebula (NGC6992/6995).

"There's no point dwelling in the dark and ignoring the light of the stars"
Carrie Hope Fletcher

The Veil Nebula, almost in its full extent - although some portions extend above and below the image

Camera/scope: Canon 300D with William Optics ZS80 and 0.8x reducer (436mm f/5.4)
Exposures: 12x300seconds @ ISO800 per panel - 3 panels to create a mosaic image

The full mosaic image also includes data blended from the following higher resolution views of the various parts of the Veil, showing strands of material knotted and interwoven. The two predominant colours are blue and red, from the ionised oxygen and hydrogen. These colours come out really well in either broadband or narrowband images.

"Those who study the stars have Deity for their teacher"
Tycho Brahe

The Network Nebula, featuring NGC 6992/6995 and IC1340

Camera/scope: Canon 300D with Skywatcher MN190 (1000mm f/5.2)
Exposures: 5x480seconds @ ISO1600

"When you reach for a star, Only angels are there, And it's not very far, Just a step on a stair"
Kate Bush

The Witches Broom, or western portion of the Veil Nebula (NGC6960) and the star 52-Cygni

Camera/scope: Canon 300D with Skywatcher MN190 (1000mm f/5.2)
Exposures: 40x300seconds @ ISO800

The bright star (52 Cygni) in the centre of the image is unrelated to the nebula and is in a coincidental line of sight. The fine tendrils of gas in the nebula are undulations in the surface of the thin shells of gas, seen edge-on by us. It is estimated that these thin shells of gas are only 4 billion miles thick – not much on the galactic scale.

"To our eyes the stars seem small; but the littleness is not the fault of the stars, but of our eyes"
Al-Maarri

Pickering's Triangle and NGC6979/6974 in narrowband hydrogen-alpha and oxygen-iii

Camera/scope: QHY23M with Takahashi E130D (430mm f/3.3)
Exposures: A bi-colour image using h-a and O-iii data (14x600seconds h-a; 10x600s O-iii)

This image only has two sets of filter data, with H-a mapped to red, O-iii mapped to blue and a synthetic green channel created by combining the H-a and O-iii data. It results in a close colour match to a standard RGB image.

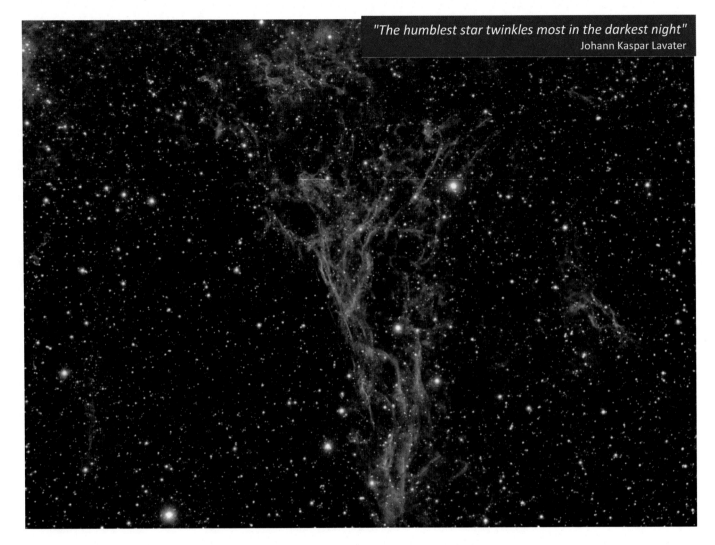

"The humblest star twinkles most in the darkest night"
Johann Kaspar Lavater

The top area of Pickering's Triangle in natural colour

Camera/scope: QHY163M with Skywatcher MN190 (1000mm f/5.2)
Exposures: LRGB image (27x300seconds; 27x60seconds per colour filter all bin 2)

The image has less contrast than the narrowband version and appears softer. The colours are remarkably similar though from this broadband spectrum image in comparison to the narrowband filtered version. Note that the amount of colour data is quite low, with only 60 second exposures. The longer Luminance exposures provide the brightness in the image.

The next supernova remnant on show is the Spaghetti Nebula, catalogued as Simeis-147, in the constellation Auriga.

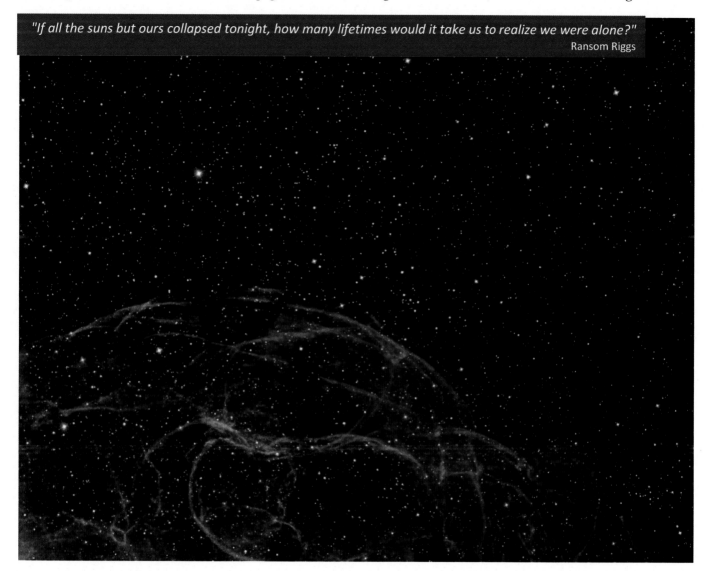

"If all the suns but ours collapsed tonight, how many lifetimes would it take us to realize we were alone?"
Ransom Riggs

The Spaghetti Nebula (Simeis-147 or Sharpless 2-247) is around 40,000 years old and spans 150 light years

Camera1/scope1: QHY23M with TSINED70mm & 0.8x reducer (336mm f/4.8)
Camera2/scope2: QHY163M with Takahashi E130D (430mm f/3.3)
Exposures (1): H-alpha; 23x600seconds [per panel]
Exposures (2): H-alpha-RGB image (12x600seconds H-a; 8x300seconds per colour filter) [per panel]

This is another large nebula spanning over 3.5 degrees. It's so large that I will need to take between 9 and 12 panels to fit into a mosaic of the whole object. So far, I've only taken 2 panels! It is a really challenging object to image well, as it is very faint and requires extremely long exposures to record the faintly glowing hydrogen gas. Each panel in this image has 5.8 hours of hydrogen-alpha data and 2 hours of colour data. So, with two panels that's a total of almost 16 hours imaging time. But the final full mosaic image will be worth the time, as this is a beautiful giant ball of gas filaments and one of my favourites. The hydrogen-alpha data has been used as both a Luminance layer and as additional data in the red channel of the image in order to make the nebula stand out more from the background sky.

In 1996 the stellar remnant was identified that was left behind after the supernova explosion. In this case, it is a rapidly rotating neutron star, or pulsar, catalogued as PSR J0538+2817. It spins once every 143 milliseconds, or roughly 7 times per second, which is amazing for an object ~20km in diameter.

A paper by Ng & Romani et al (2006) discusses the likely origin and proper motion of the pulsar and its relationship to the nebula, using data from the Very Long Baseline Array (VLBA) radio telescopes in New Mexico and the Chandra X-ray Observatory (CXO). They concluded that the pulsar is approximately 40,000 years old and that the original star is likely to have been a runaway star from the nearby cluster M36. This cluster is estimated to be 40 million years old and is an association of large O-B type stars, which are prime candidates for this type of supernova explosion.

The size, shape and estimated age of the neutron star and nebula suggest that the supernova explosion was very energetic, expanding into a low-density bubble of gas ejected by the progenitor star when it was in a Wolf-Rayet phase. This bubble of gas was likely to have been quite large, allowing the shockwave from the supernova to travel uninterrupted for a significant distance before interacting with the shell of gas, which is consistent with the size of the nebula we see today.

The likely location of the star when it erupted as a supernova is shown on the image by the cross, assuming an age of 60,000 years. If the event was 40,000 years ago then the detonation site would be just off the image further towards the "centre" of the nebula. The arrow shows the direction of the proper motion of the neutron star in space, as identified by Ng & Romani et al.

Another very interesting supernova remnant is the aptly named Jellyfish nebula, or IC443, in the constellation Gemini. The nebula can be found close by the bright triple-star system Propus (Eta-Geminorum). In this widefield shot, the jellyfish is seen swimming down in the direction of the star Tejat (Mu-Geminorum) just off image. The nebula is 5,000 light years away, spans over 70 light years and is about 30,000 years old.

"Space - the final frontier"
Gene Rodenberry

The Jellyfish Nebula (IC443) in Gemini, a complex supernova remnant interacting strongly with surrounding molecular gas clouds

Camera1/scope1: Canon 300D with TSINED70mm & 0.8x reducer (336mm f/4.8)
Camera2/scope2: QHY23M with Takahashi E130D (430mm f/3.3)
Exposures (1): RGB image (11x600seconds @ ISO400)
Exposures (2): Sulfur-Hydrogen-Oxygen image (8x600seconds S-ii; 10x600seconds H-a; 9x600seconds O-iii)

The remnant star is thought to be a pulsar, catalogued as CXOU J061705.3+222127, based on discoveries made by the Chandra X-ray observatory in 2012, although there is some doubt about that based on the orientation of the Pulsar Wind Nebula (PWNe) according to studies from 2006 and a more recent paper published by Swarz & Pavlov et al (2015). Perhaps this is a chance encounter between a neutron star and the IC443 nebula. Or perhaps it belongs to another supernova remnant nebula in the vicinity.

The 2015 paper used X-ray data to confirm that the neutron star has the characteristics of a pulsar, although no pulsing beams of radiation have actually been observed, possibly due to our line of sight. It did detect a circular ring of X-ray emission close to the neutron star and an associated jet, commonly found with pulsars, as well as a comet-shaped PWNe that is characteristic of a pulsar travelling at super- or trans-sonic speed through the interstellar medium. Although it is highly likely that the star is a pulsar, there are a lot of uncertainties in the data and it remains unclear as to whether it really is the stellar remnant that produced the IC443 nebula itself.

One of the reasons IC443 is so interesting is that the progenitor star was almost certainly still embedded in its birth nebula (a molecular cloud) when it exploded as a supernova. This is possible because B-type stars do not emit as much ultraviolet radiation as their larger O-type cousins, so they may not clear away the birth nebula in their short-lived 30 to 40 million year life. Thus, when these B-type stars, with a mass of 8-12 solar masses, go supernova they interact strongly with the surrounding birth nebula.

In the preceding image the "head" of the Jellyfish is to the bottom left (north-east) of the structure and its "tentacles" extend up to the top right (south-west) towards Eta-Geminorum. The tentacles have a very different structure, so the nebula is described as having two distinctly different shells.

The head is expanding into a low-density molecular cloud of neutral hydrogen, travelling at high speed (of the order of 100km/s) and emitting light from iron, neon, silicon and oxygen atoms that are heated by the blast and interacting shockwaves.

The tentacles, or the second shell of the supernova remnant nebula, are expanding more slowly (at around 30km/s) into a more dense and knotted area of the parental molecular cloud. These shockwave interactions are emitting light mainly from hydrogen gas energised by the expanding supernova explosion.

The dark lane crossing the centre of the Jellyfish between the head and the tentacles is the parental molecular cloud seen in the foreground, blocking out light in the visible portion of the spectrum and X-ray.

Interestingly, a third shell of gas in the area has been identified as a supernova remnant (catalogued as G189.6+3.3), but this is not associated with IC443. It is an independent nebula in the foreground. In fact, some theories point to the pulsar CXOU J061705.3+222127 as being linked to this other supernova remnant rather than IC443.

The following image utilises Sulfur-Oxygen-Hydrogen in a false colour mapping, revealing the emission nebula catalogued as Sharpless-249 as the blue area to the bottom left of the image.

The Jellyfish nebula seen in narrowband emission wavelengths of S-ii, O-iii and H-a

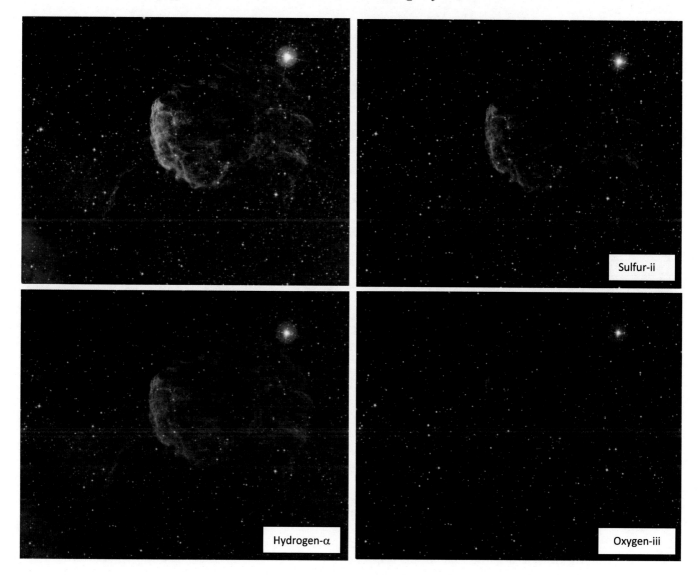

"Non est ad astra mollis e terris via"
Seneca

Close up view of the head of the Jellyfish Nebula

Camera/scope: QHY163M with Skywatcher MN190 (1000mm f/5.2)
Exposures: LRGB image (14x600seconds Luminance; 8x300seconds per colour filter all bin 2)

This image shows more detail in the complex filamentary structure of the north-east shock front as the supernova explosion crashes into the molecular cloud at high velocity (100km/s), compressing the gas and heating it up, leading to strong emission lines from hydrogen, oxygen, sulfur and other elements.

Undoubtedly, the process of stellar nucleosynthesis, creating heavier elements through the evolution of a star, and the spectacular death of larger stars through supernovae have been essential to the development of the complex mix of elements in our universe. But these events are not entirely responsible for the creation of all elements in the periodic table. The processes responsible can be summarised as follows.

- Big bang - creating hydrogen, helium and lithium
- Stellar nucleosynthesis
- Supernova nucleosynthesis
- Neutron capture
- Proton capture
- Photodisintegration from cosmic rays
- Human induced unstable elements, created through the nuclear fission of heavy elements such as uranium

In 2017, scientists detected a pulse of gravitational waves with the LIGO (Laser Interferometer Gravitational-Wave Observatory) and VIRGO, catalogued as GW170817. Current theories are that this was the result of two neutron stars merging around 130 million light years away from earth, supported by the observation of signals across the electromagnetic spectrum including visible light. These degenerate stars were 1.6 and 1.1 solar masses, which combined to produce a supermassive neutron star for a brief time, before collapsing to form a black hole.

Copious amounts of heavy elements including gold, platinum and uranium were observed to be produced by the event, the first time we have directly observed this and confirmed the theory of how the heavier elements are created.

Isn't it incredible to think that the gold in your wedding ring, or watch and the platinum in your car's catalytic converter came from such an event somewhere in our universe. Perhaps even more fundamental to us as human beings is the fact that our carbon-based life form would not exist without the nucleosynthesis process in the stars leading to the creation and balance of elements such as carbon and oxygen.

Joni Mitchell's words in her 1970 song *Woodstock* are absolutely true:

I came upon a child of god,
He was walking along the road...
We are stardust,
We are golden...
(We are) Billion year old carbon

My penultimate image for this book could be that "child of god", sometimes given the moniker the Soul Nebula due to its resemblance to a foetus. It lies in the constellation Perseus and is adjacent to another nebula, the final image in this book called the Heart Nebula, together forming the heart and soul of astrophotography.

And thus, the circle is completed with the death of the stars producing a richer and more complex environment, triggering the birth of new stars and populating the universe with incredible potential for beauty and life.

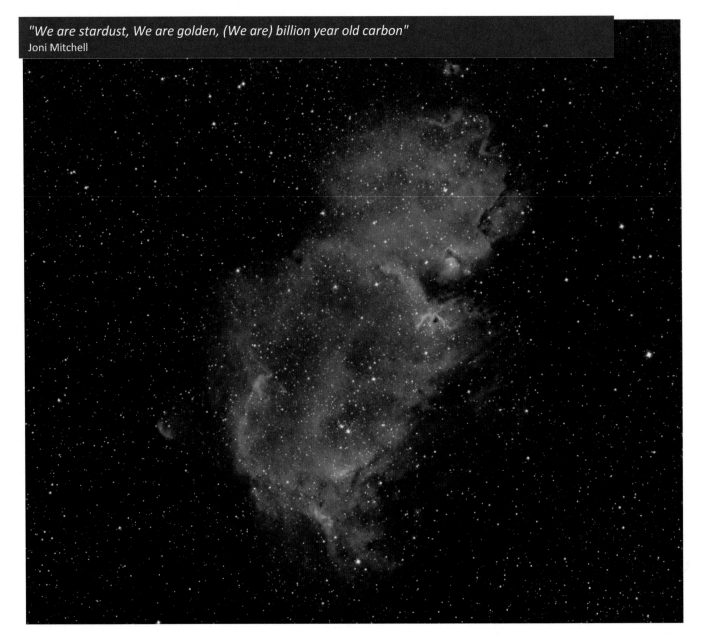

"We are stardust, We are golden, (We are) billion year old carbon"
Joni Mitchell

A 6-panel mosaic of the Soul Nebula (IC1848) in the Perseus/Cassiopeia region

Camera/scope: QHY23M with Takahashi E130D (430mm f/3.3)
Exposures: Ha-HaRGB image (8x300seconds H-alpha; 4x120seconds per colour filter bin 2) per panel

"I felt a great disturbance in the Force, as if millions of voices suddenly cried out in terror, and were suddenly silenced"
Obi Wan Kenobi

The Heart Nebula (IC1805) in the Perseus/Cassiopeia region, combining broad and narrowband data

Camera/scope: QHY23M with Takahashi E130D (430mm f/3.3) & Canon 300D with TSINED70 (436mm f/4.8)
Exposures: RGB-SHO image (22x300seconds @ ISO800; 4x600seconds per narrowband filter)

APPENDIX I
MY JOURNEY

My interest in astronomy began as a young teenager around 1987/88 and my first project was to construct my own telescope. This was a simple 6" Dobsonian telescope, made from MDF with mirrors and accessories bought from one of the few UK suppliers around at that time. I was so pleased when I was actually able to focus on the moon, and rapidly developed a desire to attach a camera to the telescope and try to replicate some of the images I drooled over in magazines like Astronomy. Needless to say, the Dobsonian was not particularly light weight and was very limited in what photography I could achieve with it. But I do remember one fine night when my parents drove us to a friend's farm to observe a lunar eclipse with my scope. And I did manage to capture the spectacle on film - unfortunately having since lost this precious picture somewhere over the past 30 years.

I soon decided that the next step was to build a tracking system to help me move into long-exposure photography. So I invested my pocket money in a book by HJP Arnold "Night Sky Photography" which gave simple tips for the absolute beginner, including an explanation of how to build a Haig Mount. For those unfamiliar with this contraption, it's a simple system that consists of two hinged boards fixed onto a baseplate, mounted at the angle of the observer's latitude. A bolt passes through the bottom board and pushes open the upper board, effectively creating a rotating equatorial axis along the hinge-line. The beauty of the mount is its absolute simplicity - it can be made in half an hour (or less) by anyone with modest woodworking skills and costs about £20-30 with a camera ball joint!

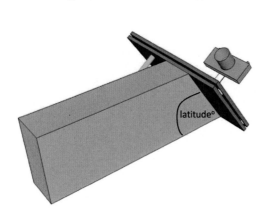
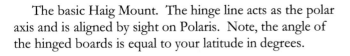
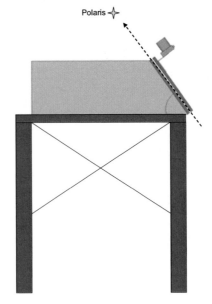

The basic Haig Mount. The hinge line acts as the polar axis and is aligned by sight on Polaris. Note, the angle of the hinged boards is equal to your latitude in degrees.

The device is limited in its accuracy, but in practice I found that wide angle lenses could tolerate up to 7-minute exposures, and 200mm lenses could go for 30 seconds before tracking errors appeared.

At the time, digital cameras were only just beginning to appear on the market, so I was using my father's 1970s Pentax Asahi SLR camera. Although this was a good solid camera, it was of its time and the results we can now achieve with digital SLRs are orders of magnitude better.

This was one of the best images I managed with my Haig Mount, taken up at the Stone Circle near Keswick in the Lake District. The exposure was 7 minutes long, taken on 400ASA film with the Pentax Asahi and a 50mm lens.

The North America nebula, Gamma-Cygni nebulosity and even a wisp of the Veil nebula are all visible in this image of the constellation Cygnus.

The dark line at the top left of the image is the aerial of my car, which I had setup next to as an additional shield from the wind!

In 1993 I went to Hawai'i on holiday with a friend and whilst we were there we constructed a Haig Mount, purchased some "hypered" Konica SRG3200 film and took the kit up to the Mauna Kea Visitor's Centre at 11,000 feet. The local guys were very accommodating, giving us the passcode to the facilities so we could make a night of it and use the toilets

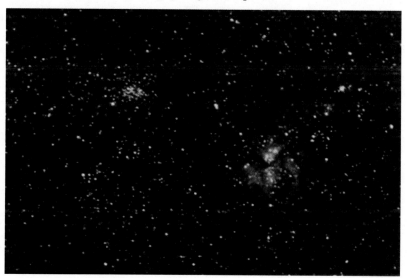

and kitchen. I can still see in my mind's eye the incredibly vivid night sky! The milky way was so bright I would swear that it almost cast shadows, and of course I could see stunning objects like the Eta Carinae nebula which were entirely hidden from me at home in the northern hemisphere.

I was so proud of this image, taken with such crude equipment knocked up in a friend's garage. Although I was already hooked, this really spurred me on to make this my lifelong passion.

Camera: Pentax Asahi SLR, 200mm f/3.5
Exposure: 1 x 30second exposure, 3200 ASA

Fortunately, technology moved on at a phenomenal pace through the 1990s and early parts of the 21st century with the advent of digital cameras and reasonably priced telescopes and mounts. For me this was also coupled with starting to earn a decent wage as a professional engineer, so it wasn't long before I ordered my first proper telescope...a 6" f/5 refractor mounted on an EQ5 mount with electric drives. It's worth reflecting on the fact that although this was a vast improvement on the home-built contraptions I'd used until that point, it was only a simple electric drive on the RA and declination axes, no Goto computers, no autoguiding.

And not long after the telescope, I bought a Canon 300D digital SLR in 2004. This revolutionised my own astrophotography.

This is an example of the type of widefield images I could take by piggybacking my camera on the telescope and manually guiding the telescope to overcome any errors in the tracking. When I think back, this was really quite hard work. Standing in the freezing cold looking into an illuminated reticle eyepiece and pressing the up/down/left/right buttons on the electric drive control pad to correct for any drift of a guide star. But that was all part of the learning curve for me and it demonstrates one astrophotography truism - persistence, patience and dedication are a must if you want to succeed.

Another typical image from the early days of simple electric motor drives, is a "prime focus" image like the following. This is taken with the lens removed from the camera and a nose piece adaptor fitted to allow the camera to be pointed directly into the telescope focuser. So, the large refractor was effectively acting as a 750mm f/5 giant telephoto lens.

At this point I wasn't able to manually drive the telescope, since the camera was pointing into the focuser. So, I relied on the accuracy of the mount and drives, which only extended to about 60 to 90 seconds, depending on how well I had polar aligned the mount. This image of the Pleiades at 750mm f/5 is a stack of 7x 1-minute exposures with the Canon 300D.

Around 2015/16 a number of modern tracking devices were released onto the market, including the IOptron SkyTracker.

These are the modern equivalent of a Haig Mount, designed to hold a camera and large lens (or perhaps small telescope), and enable exposures of several minutes to be taken. My own experience has been with the IOptron SkyTracker V2, which is very portable and has been tucked into my suitcase and taken to Italy and Sicily. The least portable component of the system is a decent tripod to mount the tracker on!

The IOptron is shown with my Canon 300D and the TS INED 70mm refractor. The camera and scope weigh just under 3kg, which is about the limit of this setup. I also use the 0.8x reducer to bring the focal length down to 336mm f/4.8.

I find the unit very easy to accurately polar align with the IPhone App that replicates the exact position of Polaris in the illuminated reticle for your location, date and time.

I have taken exposures up to 3 minutes in duration with good tracking accuracy using the 70mm telescope. And with wide angle lenses I've taken up to 11-minute exposures with no tracking issues, although I suspect longer exposures are possible.

This is an excellent way to begin astrophotography, as you don't need anything more than the Tracker unit, a tripod and your camera/lens (oh and a cable release/timer for your camera to take exposures with the "bulb" setting). It's relatively inexpensive and opens up a huge array of deep-sky targets that you can spend a long time honing your skills on. One of the most useful aspects is the portability of the unit. The less portable your system, the less likely you are to actually go out and use it, unless of course you have an observatory!

And that leads me on to another significant development in my career as an astrophotographer. In 2006 I built myself an observatory in the garden, enabling me to maximise use of the infrequent clear skies of Cumbria. Setting up the system once, getting perfect polar alignment and being able to get things running within a matter of minutes is of huge benefit.

Shortly after the observatory project, I upgraded my mount to a Skywatcher EQ6 Pro from the original EQ5. This was another quantum leap in what I was able to achieve, with a fully computerised Goto system that supports auto-guiding - and therefore unlimited exposure durations. Later equipment upgrades included imaging telescopes and dedicated astronomy CCD and CMOS cameras.

In the next chapter, I will describe a variety of simple ways to begin astrophotography, including the basics of image processing. Beginners shouldn't be put off by the thought of having to buy lots of expensive equipment, as this is not a requirement to start your journey. I'd always encourage people not to jump in and buy lots of expensive equipment, certainly until you've decided if this is really the hobby for you. Join your local astronomy society, who will often have equipment you can borrow as a member, or will certainly have lots of people with their own experiences they will be willing to share. But one thing is almost for sure, if you get hooked like I did, then the sky is quite literally the limit in terms of the equipment you will accumulate over the years! But that's probably true with most hobbies.

My EQ6 Pro has a theoretical load capacity of 18kg for imaging and 26kg for visual observing. But with careful mounting of telescopes, balancing of the weight and perhaps a little luck, I'm able to run 3 imaging scopes concurrently on the same mount with excellent pointing and tracking accuracy. That's 29kg with scopes, cameras and guidescope! Not something I'd recommend generally.

APPENDIX II
ASTROPHOTOGRAPHY - WHERE TO START

This chapter will introduce the basics of astrophotography from a practical perspective, building on the general information provided in chapter 1. First it will explain how to take images with no specialist equipment at all, then how to take images with some simple astronomical equipment, and lastly how to process those images with freely available software. Although this book focuses on deep-sky photography, I have included some information about simple lunar and planetary imaging as well since this can be a good place for beginners to start.

Untracked Images

Recalling the information in chapter 1 about how quickly the earth rotates, it is important to recognise that untracked images, taken with a camera on a fixed tripod, will be limited in how much light can be gathered before stars begin to trail across the image. Of course, the most basic image that can be taken is of star trails in a long exposure photograph.

This image was taken on the shores of lake Derwentwater in Cumbria (UK), with the Skiddaw massif in the background and the lights of the market town of Keswick providing some pretty light pollution to the lower sky.

Portraits like this of the stars rotating around the north (or south) celestial pole can be very pretty with the right foreground setting. Aim to have a good proportion of the image taken up by a nice feature such as mountains, lakes, buildings, beach scenes, etc. Depending on the light conditions of the night, you may need to light up the foreground gently with a torch. Or there may be sufficient ambient light, particularly if the exposure is quite long, not to have to add any torchlight.

Camera:	DSLR with cable release (or programmable)
Lens:	Wideangle 18mm, 28mm, 35mm
F/ratio:	Stop down a couple of stops (e.g. f/4)
ISO:	Use a low setting like 200, 400 or perhaps 800

The lower the ISO setting, the better the colours will be and the lower the camera noise will be.

Place the camera on a tripod, the sturdier the better to ensure there is no wind-shake. Set your camera to manual mode and put the exposure to "bulb" or if you can program the exposure time then set it to 30 minutes or longer. The longer the image, the longer the trails will appear, but remember that very long images will build up background "skyglow" and this may begin to drown out the stars. Also, a very long image can be ruined by things like vibration or other factors and all that time is wasted. An alternative is to take a series of shorter exposures and use software to combine the individual frames into one long exposure - various freeware packages are available online. In this case, a series of 20 second exposures would work well.

Of course, another simple image to take is a short exposure that doesn't lead to star trails but captures a nice foreground scene and some interesting stellar objects like the moon, a planet and some recognisable constellation or asterism.

The image to the right was taken on the shores of Derwentwater in the Lake District and is a single exposure with my Canon 300D, a 29mm effective focal length at f/5, ISO400 and a 20 second exposure. There is no need for tracking due to the short exposure time, so this is a simple tripod mounted picture.

The dazzlingly bright Venus can be seen beneath the Pleiades, with the V-shape of the Hyades open cluster to the left. The lake is lit up by the setting sun behind the hills of Grisedale and Catbells. This was taken on 20th April 2018 whilst hunting for the elusive aurora.

Once you have your image, it will almost certainly need some adjustment of the dark, mid and bright levels. This can be done with many software packages, but I have illustrated the basic idea below with PixInsight using the Histogram Transformation tool, which is the equivalent of the Levels tool in Photoshop.

I must say that you should always use the native RAW format of your camera (for DSLRs). Don't use any in-camera compression like JPG or TIFF, and don't use any noise suppression routines either, as these will probably blur or delete the tiny points of starlight.

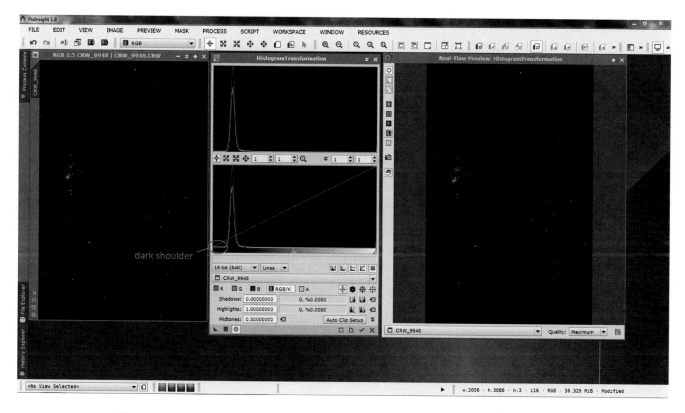

The original image of M8 (left), the HT control panel (centre) and the preview of the adjusted image

In the screenshot above, no adjustments have yet been made to the brightness histogram, so the two images are the same. The histogram shows the number of pixels in the image at different brightness levels (y-axis), with zero brightness at the left and highest brightness at the right of the x-axis.

Note the large space to the left of the histogram, sometimes referred to as the dark shoulder. This gap should be minimised so that the darkest signals in the image are closer to the "zero" brightness level, whilst ensuring there is a small gap remaining - otherwise genuine faint signals will be "clipped" or removed from the image.

There are three small triangular shaped sliders at the bottom of the control histogram, one at the left-hand end to set the "dark" level, one in the middle to set the "midtones" and one at the right-hand side to set the "brightness" levels. Pulling the brightness slider across to the left will increase the overall brightness of the image, effectively a simple linear stretch. Moving the midtones slider to the left will begin to brighten the fainter parts of the image in a non-linear stretch, which can be subtler and prevent over-exposing the brightest parts of the image.

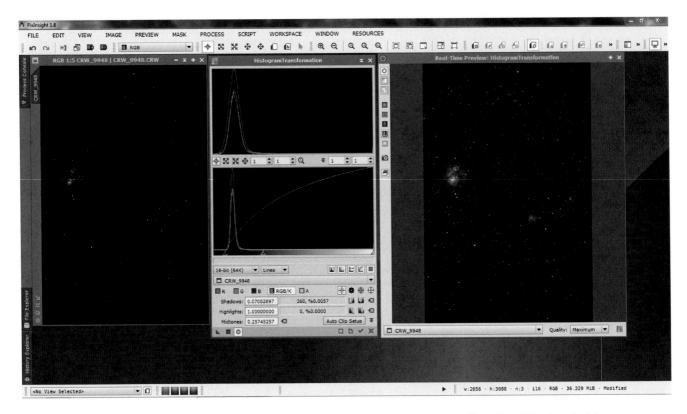

A non-linear stretch applied to the original image (left). Note the position of the "dark" and "midtone" sliders
resulting in a lot more detail in the preview image (right)

If only the midtone slider only had been used, but not the dark slider then the whole image would appear washed out with little contrast. Adjustments can be made together or one at a time as the user feels best, but the preview window is very useful in visualising the overall impact of adjustments on the image.

This principle of stretching the image is one of the most basic concepts in image processing. It needs to be carefully applied, as stretching the image too much will produce a noisy unpleasant image, whilst not stretching the image enough will result in faint details not being revealed.

A common trap for beginners to fall into is clipping the dark signal in a desire to achieve a really "black" sky background. If the brightness histogram starts at the very left-hand edge of the histogram then the user has been too aggressive using the dark slider.

A second type of untracked image that can be taken with no specialist equipment is quite astonishing in my opinion. I deliberately ran this test when I went to South Africa on a short trip in 2015, to see how much detail could be revealed in a deep-sky object using nothing more than a camera, long focal length lens and a tripod.

Recall the "500 rule", whereby the maximum exposure that can be taken (in seconds) is 500 divided by the focal length of the lens in mm. So, with my 200mm lens and a form factor of 1.6x for the Canon 300D that's 320mm, giving a maximum of 1.5 seconds before the stars begin to trail. There are some factors that can influence the rule, so I decided to push it a bit and see if I could get away with 4 second exposures. As you can see in the image, there is a little bit of trailing evident in the image, but it's broadly acceptable for this trial. I determined that 4 seconds was as far as I wanted to push it simply by taking a series of exposures at 2, 3, 4 and 5 seconds and seeing which one I could tolerate! That's the beauty of digital imaging, try it and if it isn't working adjust what you're doing until it works.

Below is a stretched (levels adjusted) single frame to show how poor the result would be if we only took one picture.

Eta Carinae nebula, Canon 300D with 200mm f/ 5.6, a single 4 second exposure @ ISO1600

The level of noise in the image is quite significant as there is not much signal in such a short exposure and the level of amplification is quite high with an ISO setting of 1600 on this camera. So how can we combat this signal to noise problem? Recalling the lessons from Chapter 1, we can significantly improve the signal:noise ratio by taking a series of images and stacking them. So, I took 26 images of 4 seconds each and used the freely available stacking software DeepSkyStacker to align and stack the images to produce a final master image that could be stretched.

Eta Carinae nebula, Canon 300D with 200mm f/5.6, 26 x 4 seconds @ ISO1600

It's rather self-evident that the amount of detail now present in the image is incomparable to the single exposure. I was quite taken aback that such a result could be obtained with a simple camera and tripod, with no tracking equipment required. This image would also be significantly better with a more modern DSLR than my old 300D!

Compare this image with the one on page 85, which was a single 30 second tracked exposure from 1993.

The Aurora Borealis

Another beautiful event that can be photographed very easily is the aurora. Of course, the tricky bit is being lucky enough to have clear skies when there is a storm on, helped by living as far north as possible. Modern aurora predictions are pretty accurate, using data streamed from solar probes and magnetic detectors on earth, so it's pretty easy to know if an aurora is occurring now compared to say 20 years ago. Generally speaking, the aurora can be visible from anywhere in Great Britain, but it's typically visible at least two or three times a year from the north of England.

Each storm is different, some being stronger and brighter than others, but typically an exposure time of between 5 and 20 seconds is about right with a modern DSLR, using ISO400-800. The longer the exposure, the brighter it will appear, but you may get some blurring of fine structure in the beams of light if they are moving around quickly. No tracking is necessary, just a camera on a tripod. Below is a single 60 second image at ISO400 with my Canon 300D.

Tracked Images

Before we look at the use of the basic image stacking software, let's take a look at some more sophisticated ways of getting more data into our images. Namely, the use of some kind of simple tracking device - this could be either a homemade Haig Tracker or a proprietary platform such as the IOptron SkyTracker (or other device) shown in Appendix I. So first of all, a few practical tips on using the IOptron. The maximum payload for this system is about 3.5kg, which equates to a small light-weight refractor or a large telephoto lens plus your camera. The unit can track in either north or south hemisphere, so it's suitable to go on your holiday, particularly as the unit itself only weighs 1.1kg. I would recommend setting the system up indoors unless you are very familiar with it, as follows:

- Insert fresh AA batteries (lasts 24h) and put some spare camera batteries and IOptron batteries in your bag
- Slot the polar scope into the IOptron body
- Attach a ball mount to the mounting block, this enables you to point your camera in the desired direction
- Set up the tripod and attach the IOptron
- Attach the camera to the ball mount
- Connect a cable release to the camera so you don't have to press the camera button with your finger

When it's nearly dark take your system outside and allow the camera and lens/scope to cool down for 30-60 minutes. The longer the system cools down the better. Then proceed to align the IOptron as follows:

- Place the system on hard ground (not decking or soft mud)
- With the IOptron vertical rotate the tripod so the system is pointing North
- Adjust the tripod legs to get the system level using the tripod bubble
- Tilt the IOptron using the latitude knob so that the polar scope is pointing at Polaris
- Turn on the illuminated reticule
- While looking through the polarscope make small adjustments to the latitude knob (tilt) and the azimuth (left/right) so that Polaris is centred on the green crosshair (if you are using the IPhone App)
- Lock the latitude and azimuth knobs securely

When you adjust your ball/socket joint to point your camera at the target you want to image, the shifting weight of the camera can cause small errors in your polar alignment. So, get your camera pointing at your target before doing your final adjustments. It's worth checking your polar alignment if you make any adjustments to the camera direction. Adjustments to focus and camera settings will not substantially alter your polar alignment, but if you point the camera in a new direction always check and adjust alignment as it only takes a few seconds once you get used to it. Now take your images:

- Focus the camera as accurately as possible, using your live-view or playback mode to check focus
- Set the camera to Bulb and set the ISO to between 400 and 3200 (the higher the more sensitive, but noisier)
- Check the f/ ratio of the lens and adjust to almost wide open (e.g. f/2 rather than f/16)
- Take a series of images (minimum 10) of the same object, one after the other

Some general tips on improving the quality of your images:

- The darker the observing site the longer your exposures can be – avoid areas with street lights
- Generally speaking, the longer your exposure the better the signal:noise ratio will be. Shorter exposures have more noise. Recognise the limits of your tracking accuracy though - the longer the exposure the more likely you will have to discard images that have trailed due to tracking errors/vibration.
- Try to find shelter if there is a breeze/wind as this will vibrate the image and cause trailing particularly if you are using a longer focal length lens or small scope
- Use a lens shade to minimise dew accumulating on the lens – keep checking this, particularly in autumn. If you get dew on the lens it may be possible to wipe it off with a lens cloth, but if you can use a hairdryer that's better as you won't smear the lens or affect the focus
- Remember that increasing the ISO setting is just turning up the gain factor in the camera. It doesn't make it inherently more sensitive. So, there is a compromise between boosting the signal by increasing ISO and reducing the noise levels by reducing ISO. It's something you just have to experiment with for your camera.

The following images were taken using the IOptron, with my Canon 300D and either a standard camera lens or the TS INED70mm telescope. They were taken in Sicily and Italy during vacations. I have processed the images using PixInsight and Photoshop, to create the best images possible, but they could easily be processed in with the free software DeepSkyStacker.

The Lagoon Nebula (M8) and the Triffid Nebula (M20) taken from Italy in 2016 with Canon 300D and TS INED70 with 0.8x (336mm f/4.8) using an IOptron Sky Tracker V2 Exposure time is 10x60s @ ISO1600

This image is taken with a wide-angle lens of 29mm. The camera is mounted on the IOptron tracker and this is a single exposure of 11 minutes duration at ISO400.

It shows the Andromeda Galaxy (M31) as a ... all smudge at the bottom of the image rising over a castle in Italy. Overhead, the milky way can be seen, with the bright red nebulae in the north of Cygnus, including the North America nebula and the Gamma-Cygni nebulosity. Below these, the red nebula IC1396 can be seen just above the red/yellow coloured Garnet Star in Cepheus.

The foreground trees and castle have blurred slightly as the exposure was for 11 minutes and the tracker rotated to follow the stars, with the land remaining still.

But very often we want to take lots of images and stack them together to get a much cleaner signal, rather than just one long exposure. Let's take a look now at the process of aligning and stacking images using the free software DeepSkyStacker. This is only one tool, but I've chosen to demonstrate it in this chapter for two reasons. Firstly, it's freely available to download from the internet, so it's ideal for beginners to use without having to invest hundreds of pounds. And secondly, it's very simple to use but has some powerful features. Of course, it has its limitations and it's not something I would use for more serious work. But it's a great place to start.

Image Calibration

Before we begin with the detail of how to use the software, it's important to discuss the principles of image calibration. On page 10 of Chapter 1 we briefly introduced 3 main sources of noise that arise in our data, so let's recap what those are and how to combat them.

(1) The Poisson noise associated with the quantum nature of light, which is random;
- This is counteracted by taking multiple images and stacking them together with a statistical process (averaging).
(2) Thermal noise generated by the image sensor itself, sometimes called the dark current of the pixels;
- This is counteracted by taking a "dark frame" - essentially a picture of equal duration to the main image, but with the lens cap on. In other words, it's a picture of nothing but the thermal noise in the sensor itself. The dark frame also corrects for hot pixels in the sensor.
(3) Read noise generated by the electronics in the sensor
- This is counteracted by taking the shortest exposure possible (with the lens cap on) to capture only the electronic noise generated during the process of reading the data from the CMOS/CCD chip array and converting it into a digital signal. This is called a "bias frame".

In addition to these, there is a final correction, or calibration that can be applied to the raw image data. This is called a "flat frame" and its purpose is to correct for optical distortions such as vignetting (the uneven illumination of the chip by the light cone) and the presence of dust on the sensor (which reveals itself as circular rings of uneven brightness). Before aligning or stacking our images, the processing software is going to calibrate every "light" frame that we've taken, using 3 types of calibration files. But how do we create these calibration files in the first place?

Creating a dark frame:

- Ensure the lens cap is on the camera so no light can enter.
- Set the ISO to the same as that used for the main image
- Try to take the dark frame at the same temperature as the main image (don't go inside to a warm room if you've taken the images outside in the freezing cold)
- Take a series of exposures of the same duration as the main image - minimum 3, preferably 10 or more
- In DeepSkyStacker these dark frames can be loaded directly into the software, whereas in other programs like PixInsight or ImagesPlus, a master dark calibration file must be created by averaging the dark frames.

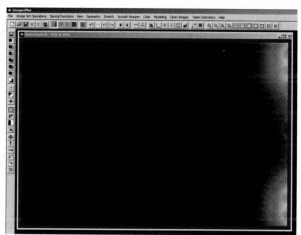

The bright lobes at the right-hand side are caused by thermal glow from the CMOS amplifier in my Canon 300D over long exposures. This is less of an issue with modern DSLRs.

Creating a bias frame:

- Ensure the lens cap is on the camera so no light can enter.
- Set the ISO sensitivity to 100 or 200
- Take the shortest possible exposure on your camera
- Repeat this for a minimum of 3 frames, preferably 10 or more
- Individual bias frames loaded directly into DeepSkyStacker.

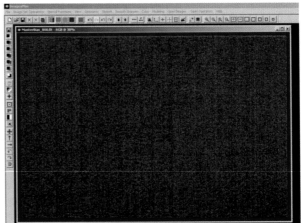

Note the vertical noise banding in this stretched bias frame image, which will be corrected out of the final light frames.

Creating a flat frame:

This is trickier, as the intent is to have an evenly illuminated light source but with no features in the image. So the lens cap cannot be on, but equally we don't want to record any objects like stars or background objects in the image. There are two ways to do this effectively in my opinion. Either, buy a flat fame box, which is a light source generator with a gauze like cover that diffuses the light evenly. This is placed over the front of the lens or telescope. Or, point the lens at a blank patch of white or grey board which is evenly illuminated with no shadows.

- Set the ISO to around 200 or 400
- Set your DSLR to aperture priority; take 10 or more images.

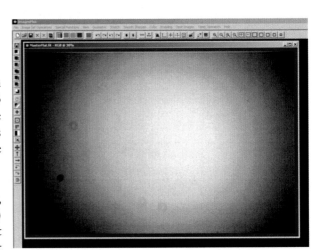

The secret is to expose the image so that the maximum brightness in the image histogram is about 2/3 of the maximum. Don't over-expose it or you will lose the subtlety of darker areas at the corners of the image where there is vignetting. By using aperture priority, the camera should give you the right exposure by default.

The position of all aspects of the optical train (lens, focus point, any filters, the orientation of the camera with respect to the lens) all need to be exactly the same for the flat frame as for the light frames. Take the flats immediately after finishing your main light exposures. Otherwise the position of the dust motes will be different and you will incorrectly calibrate for them.

For DSLRs dark frames should really be taken for each imaging session. Some people build up a bank of dark frames and use the ones they feel match the atmospheric temperature best. Of course, with temperature-controlled CCD or CMOS cameras, a pre-made bank of dark files is perfectly acceptable. But not so with DSLRs.

The bias frames tend not to change much with time, so a bank of pre-made bias frames is fine.

Using DeepSkyStacker

One thing to note before we look at the DSS workflow is to be organised and tidy with your filing system or you can easily become muddled with your images. I always create a new folder for each object that I image, titled with details of the object and the imaging system. Within the main folder, create subfolders for lights, darks and flats (assuming that you already have a library of bias files somewhere else). And I usually create a temporary folder within the main folder for image processing files.

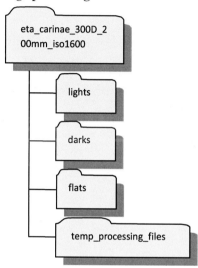

The user interface for DSS is very intuitive, following a logical order of steps. This tutorial is based on DSS v3.3.2.

Registering and stacking is the main feature of this program. By registering, we mean aligning all of the image frames so that the stars are in the same position. No matter how good your tracking, there will be minor, or possibly major, differences between the image files, making this an essential step. Once the image files are all registered, the program will then move to the stacking process, where all of the aligned images are merged using a statistical process to improve the quality of the final image.

The first step is to load the "picture files" or the light frames of the object that you have taken.

Click *"Open picture files..."*

select the correct folder with your light frames in;

highlight them all and *"Open"*

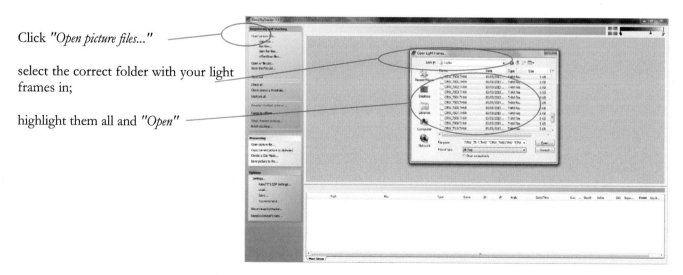

Once loaded, click "Check all" to select all of the files and select click one of the file names to see a preview of the image.

The screen will look like this:

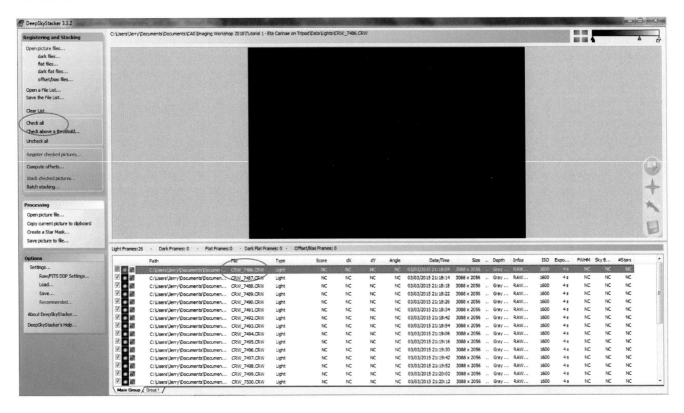

Next, load the dark files in the same manner. And follow with the flat files. And then the bias files.

The software gives an option of dark flat files too, but ignore this. In fact, you can ignore any of the calibration file types if you don't have them. But of course, it's best to ensure you do have darks and flats as an absolute minimum to ensure you image is as clean as possible.

You will need to scroll down to see the dark, flat and bias files once they are loaded. You can click on any of the files in the list to see a preview of it. It's worth quickly doing this to ensure that you've only imported valid files.

Note the details of each file are logged by the program automatically including date, time, ISO and exposure.

Next, click *"Register checked pictures..."* and you will see the following screen.

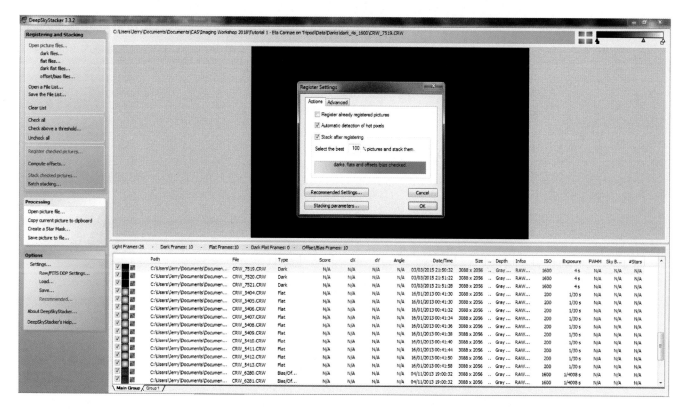

Click on the option Stacking Parameters in the dialogue box that appears. This will open up some options for the user to select. But it's just as well to go with the standard options for starters and then go back and try different options later to see if you can get a better result.

The software has some good in-built processes like hot-pixel removal, which can be controlled with certain parameters, as well as a variety of statistical methods to use for stacking. Some work better than others depending on the number of light frames you have. Options which include "clipping" will reject outliers, so if there is a frame with an aeroplane trail in it, the trail of light will be clipped out or removed if you use the clipping option.

Once you're happy with all of the settings click OK and the process will proceed automatically, finishing by presenting you with your final image.

You can now adjust the image using some of the "Processing" options in DSS if you like. But I find this functionality very clunky in DSS, so I would simply save the image at this point and import it into another image processing program such as Photoshop or whatever you use. The post processing steps will be discussed next on the basis of using Photoshop, but the principles apply in many pieces of software. Save the image as a 16-bit TIFF.

Simple post-processing

The adjustment of the image after stacking is generally referred to as post-processing. At this stage, the image is likely to be quite dark and may not have the right colour balance. We can also carry out smoothing and/or sharpening routines to reduce the noise or enhance detail in the image. A straightforward workflow that should be followed for most images is as follows.

- Adjust the **Levels** in the image, as per the details on pages 91 and 92, using the midtones and blackness sliders. This can be done iteratively rather than all in one go in order to control the process more carefully.
- Adjust the **Colour Balance** if necessary, working on the Shadows, Midtones and Highlights separately. Alternatively, use the **Selective Colour** tool to adjust the hue of each individual colour channel.

This is an image of the Horsehead nebula processed with DeepSkyStacker and simply given a Levels and Colour Balance adjustment in Photoshop. The image is 27x 10-minute exposures at ISO800 with the Canon 300D and an 8" f/5 telescope.

A worthwhile investment for novice astrophotographers is Noel Carboni's Astronomy Tools action set for Photoshop. It only works on full versions, not limited ones such as Elements. It is around $21 and has a good range of prepared actions that can help enhance your image, without any detailed knowledge of how to use Photoshop. Of course, over time you will learn to use some of the tools yourself in a more controlled way, but it's a good start.

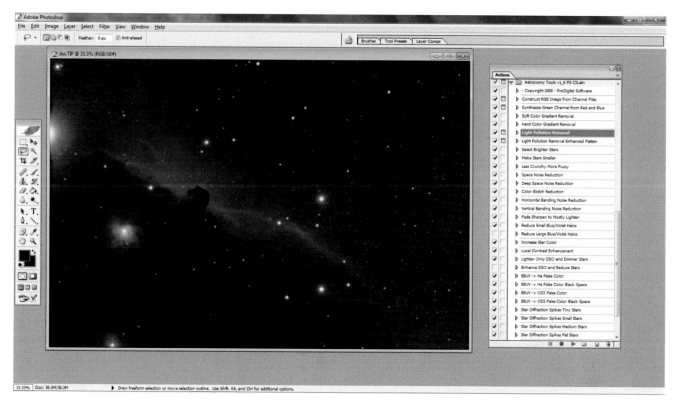

Some of my favourite routines in this action set are Vertical Banding Removal (for DSLRs particularly if you use a simple stacking tool like DeepSkyStacker), Increase Star Colour, Space/Deep Space Noise Reduction and Make Stars Smaller.

One tip: Before you apply any action, make a copy of the layer by pressing CTRL-J, open the Layer view panel by pressing F7 and then apply the action to the top layer (automatically named Layer 1). When the action is complete, you can toggle Layer 1 on/off with the Visibility icon and see what impact the action had on the original base layer. If you think the action was too strong, you can reduce it by changing the Opacity of Layer 1 and then blend the two layers by selecting Layer-MergeVisible (SHFT-CTRL-E).

Always ensure you save the original version of your image and work on a copy in case you mess it up!

Lunar / Planetary Imaging

Although this book really focuses on deep-sky astrophotography, I thought I would include a brief section on how to begin to tackle lunar or planetary imaging as this is actually one of the simplest areas for a beginner. The process is quite different in many respects to deep-sky work and requires different equipment. However, there are more cameras available on the market now that can do both types of imaging. And if you only wish to invest a modest sum of money in a camera then lunar/planetary cameras can be bought for only a couple of hundred pounds.

You can attempt to use a DSLR for this type of imaging, but the results will be quite severely limited, perhaps with the exception of some lucky shots of the moon that can be had with a DSLR. Let's see why this is.

The most important concept in this field of astrophotography is that of stacking lots of images to significantly improve the signal to noise ratio and to capture those fleeting moments of perfect seeing when the atmosphere is momentarily still. One only has to look through a telescope at the moon or a planet to see visually how much the object "bounces around" in the field of view, some nights more so than others. This is simply the movement of our atmosphere, changing the way the light is refracted as it passes down to the earth's surface. This is much more apparent in lunar and planetary images because the objects are extended objects. In other words, they have a real "surface area" that we are observing, rather than just appearing as points of light in the far distance of photographic infinity like the stars.

So first of all, let's think about what equipment we will need (and why):

- A camera with a high frame-rate output.
  Effectively this is a webcam or video camera. We want to capture something of the order of 50 frames per second or more. If we then take a video clip that lasts 2 minutes, we would have 6000 frames in the video. Each frame can then be extracted from the video file as an image, which can be aligned and stacked to give a very clean image, with the rubbish ones discarded.

- A laptop to connect the camera to so you can stream the video via USB in real time.

- A tracking mount of some kind.
  An alt-azimuth mount with a simple electric motor is sufficient as long as it will keep the object in the field of view for several minutes. Inaccuracies in tracking are much less relevant as the software will correct for the movement of the object over quite large distances.

- A long focal length lens, or a telescope for planetary work.
  This is because the planets will only ever have a maximum size of about 50 arc-seconds (for Jupiter). And most will be smaller than this. The following images produced by Ron Wodaski's CCD calculator software illustrate the point.

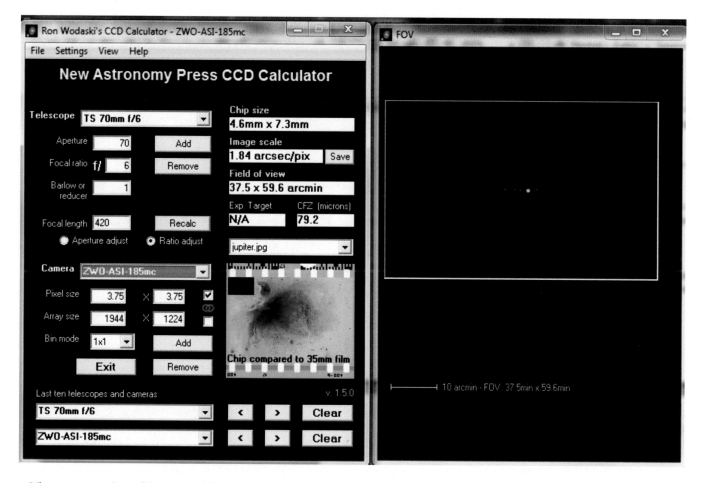

The camera selected here is a ZWO ASI-185mc, which is a high frame rate planetary camera (cost around £300). The telescope selected is a 70mm f/6 refractor with a focal length of 420mm. So, this is the equivalent of a pretty long focal length camera lens. Even with this dedicated camera, which has a small sensor that magnifies the field of view, and a fairly decent telescope, Jupiter is pretty small in the overall field of view (which is about 1 deg x 0.37 deg).

This is why planetary imaging is often done with Schmidt-Cassegrain type telescopes, or other catadioptric scopes with much longer native focal lengths of 2000mm or more. Although these scope have slower focal ratios, the camera gain can be boosted to account for the fainter image. And in practice, a barlow lens of some kind is also often used to boost the focal length by a factor of 2x, 3x or even 5x. An example of this is shown next to illustrate the size of the planetary image with such a setup.

The field of view with a 9.25" f/10 SCT with a 2x barlow and the same camera is now only 3 arc-minutes x 5 arc-minutes. So, Jupiter will be much larger in the image, with a lot more surface detail visible. If a 5x barlow was used the vertical would be only 78 arc-seconds, so Jupiter would only just fit on the screen! Of course, with such high magnification the tracking does become more important. And the image does become a lot fainter. So as with everything, begin with the simpler challenge of not so much magnification and learn from there.

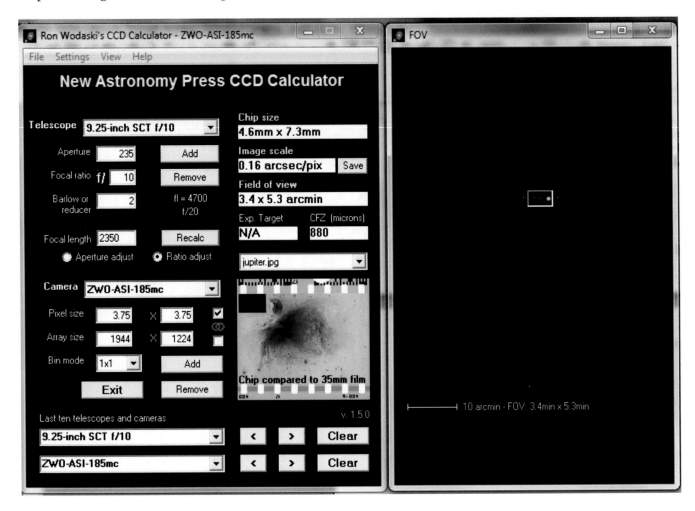

Once you've managed to capture a movie file of around 2 minutes, this can be imported into the fantastic free software packages available for planetary imaging such as Registax, Autostakkert2 and PIPP.

Some of the challenging aspects of planetary imaging include finding the object and centring it in such a small field of view; and accurately focusing on the planet.

A much simpler target to practice your technique on is the moon. It's almost impossible to miss in terms of alignment even for the absolute novice. And focusing is much easier as it is so much brighter than the small planets. A fun challenge when you have mastered the basics of lunar imaging is to take a series of close-up images and combine them into a high-resolution mosaic of the moon like the one below, comprising 8 panels.

Lunar mosaic taken with Skywatcher MN190 and QHY5 camera

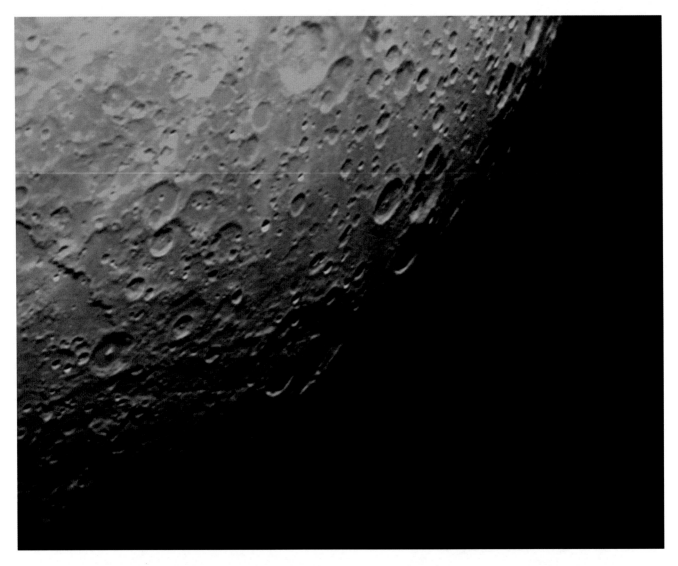

A close up view of the lunar surface, taken with a 2.5x barlow, the Skywatcher MN190 and QHY5 camera. The effective focal length here is 2500mm f/13. This is an image stack of 600 frames using Registax.

To finish this chapter are a couple of images of Jupiter. The first is taken with the QHY5 mono camera and the second is taken with a colour webcam, the Toucam Pro 2. Amazing what simple equipment can achieve.

Jupiter and some of its moons, taken with QHY5, Skywatcher MN190 and 2.5x barlow using the best 50% of 1200 frames stacked in Registax

Jupiter and its moon Ganymede, taken with a Toucam Pro 2 webcam, Skywatcher MN190 and 2.5x barlow using the best 50% of 600 frames stacked with PIPP and Autostakkert2

I readily acknowledge that my lunar and planetary imaging skills need some significant development, but I'm content to dabble in the field while I concentrate on my real love - deep sky!

APPENDIX III
ASTROPHOTOGRAPHY - ADVANCED IMAGING TECHNIQUES

In this chapter I want to go a little deeper into some of the image capture and image processing techniques that I use. You can choose to ignore these initially, or take from them what you will as early in your journey as you want. Sometimes people can be overwhelmed by trying to do too much too soon, so my encouragement would be to practice the basics and when you feel confident then keep pushing to do more.

I will talk more in this chapter about processes you can apply in Photoshop, ImagesPlus and PixInsight. I will assume a certain level of familiarity with the software, so if you don't understand certain terms you'll have to do some homework on the internet. As an aside, in my opinion, as you progress to more advanced work you will certainly have to invest in at least one of those packages. My recommendation would be to buy PixInsight - it is one of the most comprehensive and powerful pieces of astronomy imaging software you can buy, at a fairly reasonable price in terms of commercially available software. It is quite complex, but it can be used relatively simplistically as well. There are many fantastic blogs, videos and tutorials online and of course you can buy books on how to get the most from the software. I do find that some processing routines are just a little easier or better in one piece of software or another - it's not to say that one package couldn't do everything, but I guess you get a feel for what you prefer. I commonly use all three packages to process a single image at different stages, as I'll explain.

High Dynamic Range Images

In several of the images presented in the book I have mentioned the idea of producing images that have a high dynamic range. Essentially this means that the object has a wide range of brightness, from very bright areas to very faint areas. And care needs to be taken not to lose detail in the bright areas by over-exposing them when trying to capture details in the fainter parts of the object. A couple of examples of objects that present this problem are the Andromeda Galaxy (M31) and the Great Orion Nebula (M42/43) - refer to pages 6, 9, 14, 17, 44.

If you took a set of exposures of either of these objects at around 300s or longer, you would certainly overexpose the brighter parts of the object. So, the initial task is to take a series of images of different durations. Some short ones to use for the brightest parts and some long ones for the fainter areas of the image. Of course, you can take as many sets of images as you like, it doesn't have to just be a short and a long one. But I'll assume for now that we are only going to take two sets. As a case study, let's assume we've taken two sets of data of M42 as follows:

- QHY23 mono CCD camera with TS INED 70 and 0.8x reducer (336mm f/4.8)
- Set 1: 10 x 300 seconds & Set 2: 8 x 60 seconds

Each exposure set will be treated as a separate image that we create. The two images will then be combined using a process called Layer Masks in Photoshop. Of course, if you have a modern version of Photoshop or PixInsight you can use one of the automatic routines to create HDR images instead. Below are the aligned and stacked images created from the two data sets.

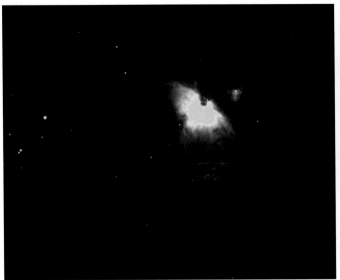

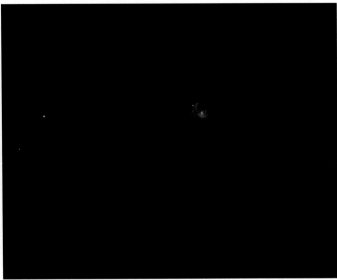

10 x 300 seconds   12 x 10 seconds

It's obvious how overexposed the core of the nebula is in the 5-minute exposures, as well as how much of the outer portions of the nebula are missing in the 10 second exposures. Neither on its own is any good really. So how do we combine the two to make a better image?

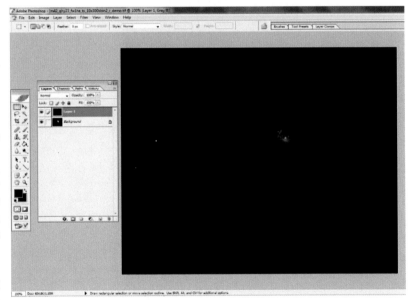

- Open both images in Photoshop at the same time. Then select the faint image (CTRL-A) and copy it (CTRL-C). Close that image and switch to the brighter image. And finally paste the faint image in as a separate layer (CTRL-V). If you press F7 you can open the Layer window, which you will need. The screen should then look like this.

- Next click the little icon in the layer control panel which looks like a square with a white circle inside. Or go to the menu at the top of the screen and select Layer - Add Layer Mask - Reveal All. A white box appears in the control panel line for Layer 1, which is the mask layer.

- We now paste the bright layer into the mask layer as follows.

    o Click on the Background image (the bright one) in the layer control panel.
    o Press CTRL-A, followed by CTRL-C to select and copy the image.
    o Whilst pressing the ALT button on the keyboard, click on the white box in the layer control panel to edit the Layer Mask in Layer 1. The screen should now go white.
    o And finally, press CTRL-V to paste the bright image into the layer mask. The screen will now look like this.

- Now go to the Window menu and select Window-Arrange-New Window. This will open a second window to visualise the changes that we will make in the original window. So just arrange the two windows so you can work in one and see the impact of your changes as a preview in the other.

- Next, in the original window, go to Filter-Blur-Gaussian Blur and a control panel will open. Play with the slider on the pixel radius until the preview looks like the blend between the bright and faint images is correctly balanced. In this instance, a setting of 10 pixels looks about right, but this will vary with the image you are processing. Press OK to enact the blur.

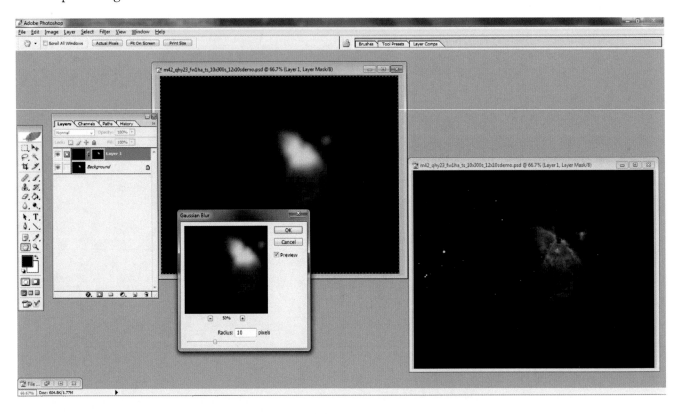

- Now we can adjust the aggressiveness of the blend by changing the contrast in the blurred mask. To do this, open the Curves tool by selecting Image-Adjust-Curves. Again, play with the settings until you get a pleasing result.

- Finally, through the menu select Layer-Flatten Image and then close the preview window as you don't need this anymore. From this point you have a well-blended HDR image that can be worked on further. Save it at this point. With this example, the image now looks like the following.

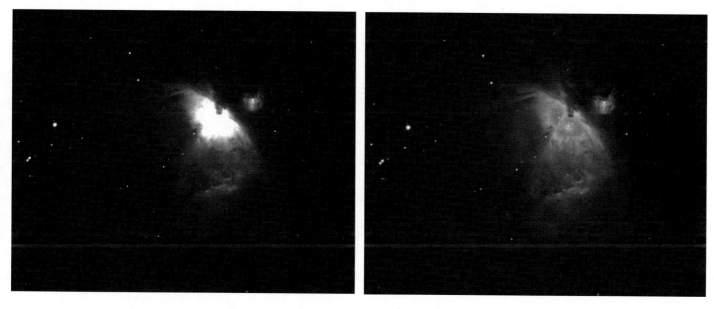

Original "bright image": 10 x 300s   Final HDR image: 10 x 300s & 12 x 10s

The original "bright image" with the overexposed core is shown in comparison to the final HDR image. I've also applied a slight stretch to the final HDR image to bring a bit more of the outer parts of the nebula to life. Obviously, there are lots more steps we could undertake next to refine the image further, but in terms of having an image which covers a wide range of brightness levels, this is the aim of HDR processing. The intensely bright Trapezium stars that light up the gas are now clearly visible in the core of the nebula.

Creating a Colour Image In Photoshop

This next tutorial will show how to create a full colour image from three monochromatic frames using Photoshop. This can be done automatically in programs like ImagesPlus and PixInsight, but I prefer the level of control that comes with doing it in Photoshop, particularly when I'm using narrowband data. This tutorial will use narrowband data (h-alpha, O-iii and S-ii) but the principle is exactly the same for red/green/blue images.

The start point for this tutorial assumes that the individual images have been calibrated, aligned and stacked to produce a single image for each filter; and that each image has been stretched appropriately. The images are then loaded up in Photoshop and copied/pasted as separate layers into one image file (in the same way we did with the HDR image creation above). So, at this point you should have a single image open, which contains 3 layers as follows. Save this as a .PSD Photoshop file.

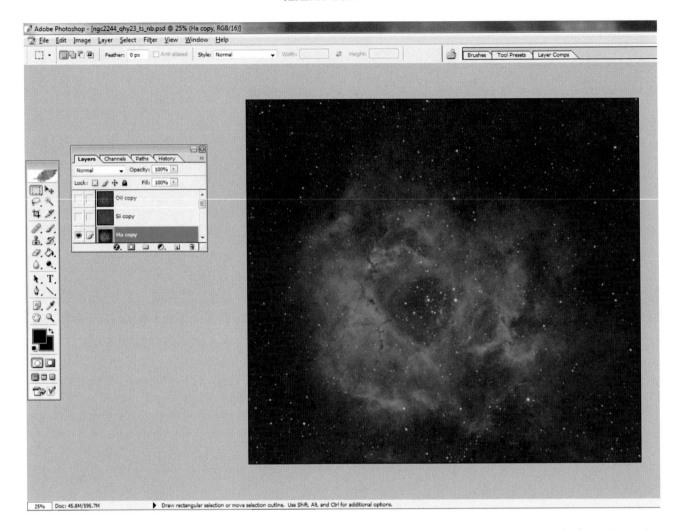

- Turn off the top two layers, so that only one layer is visible. And change the blending mode from Normal to Screen. [Note: When all of the layers are turned on using Screen, it's like projecting each layer onto the wall together resulting in them being blended together to make a final image.]
- For narrowband images you need to decide which filter you will assign to which colour channel. The standard Hubble palette is SHO, where sulfur is mapped to red, hydrogen to green and oxygen to blue. But the choice is yours. In this demonstration I'm going to use the SHO palette to get the classic gold and blue image often seen in Hubble images.
- First of all, in the layer control panel, select the S-ii layer and then click on the Create New Adjustment Layer (the icon is the circle with half black-half white fill) - and then select Hue/Saturation.

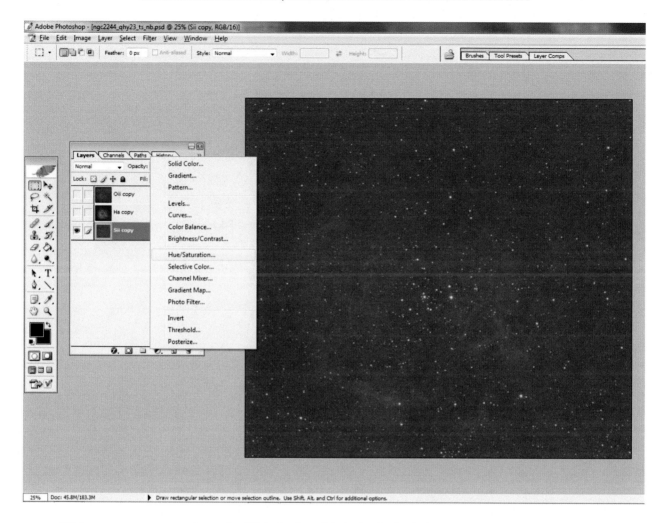

- This opens up a control panel. Click the box "Colorize". And then using the slider or type in the boxes the following numbers: Hue = 0 (red); Saturation = 100%; Lightness = -40 and then click OK. Then press CTRL-G to apply that adjustment layer to the layer below only. The image will now look red.
- Next, select the H-a layer in the layer control panel and ensure its blend mode is set to Screen. And follow the same process to create a new adjustment layer as we just did for S-ii. This time though the settings need to be as follows. Hue = 120 (green); Saturation = 100%; Lightness = -40. And again, press CTRL-G to apply the green adjustment to the H-a layer only. The image will now look a strange orange-green colour.
- And finally, do the same with the O-iii layer, using the following settings: Hue = 240; Saturation = 100% and Lightness = -40. Press CTRL-G again. The image on screen should now look like this.

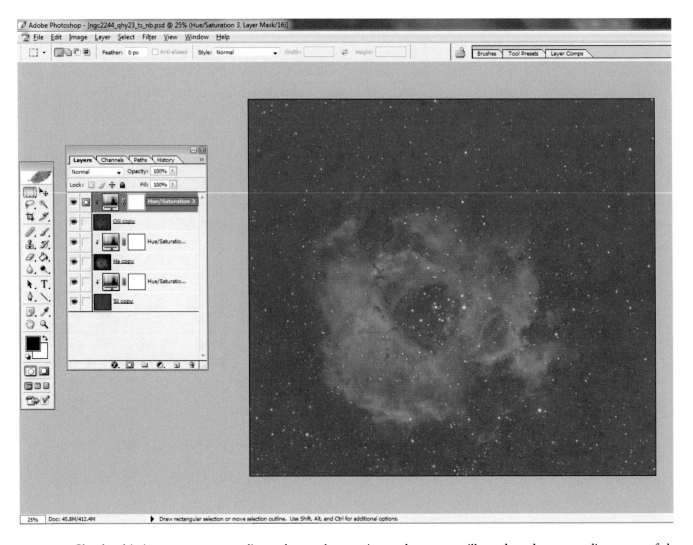

Clearly, this is not a very appealing colour palette as it stands, so we will need to do some adjustment of the various colour channels using the Selective Colour tool.

- Before you do any more alterations to the image, save it as a .PSD Photoshop file.
- Next, blend all of the layers together into a single layer by selecting Layer-Flatten Image.
- The first thing we are going to do to this colour image is a quick adjustment to the brightness levels by selecting Image-Adjust-Levels (or pressing CTRL-L). Adjust the shadows, and midtones sliders so that the image has a decent level of contrast, remembering not to clip any of the shadows.

- Now select Image-Adjust-Selective Color to open a colour control panel. This tool allows the user to change the colour balance of individual colour channels (Red, Yellow, Green, Cyan, Blue, Magenta, White, Neutrals and Blacks), which is much more powerful than just working on red, green and blue. The control panel looks like this.

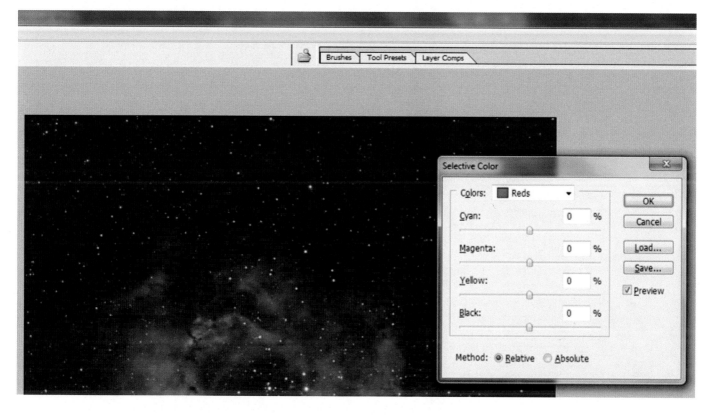

- Again, you can make whatever adjustments you think are best to get the colour balance you want. But the way to bring out the classic golds and blues in an SHO image is as follows.
- Green - this needs to become more yellow, do this by adding red, perhaps reducing magenta and adding yellow
- Yellow - this needs to become more red and less green
- Cyan - this needs to brightened, with more cyan and more blue
- Blue - needs to be brightened, with more cyan and more blue
- Reds - need to become redder, more yellow and less magenta
- Blacks and neutrals may need to be balanced to give a more neutral colour and be darkened a little

The process should be iterated until a pleasing colour balance is achieved. This is very much personal choice.

The Rosette Nebula, taken with S-ii, H-a and O-iii data, mapped to RGB respectively and then selectively colour balanced as above

Selectively Reducing Noise

The image below is a crop from the one we just created in the previous tutorial. There is some unwanted noise in the fainter parts of the image, noting that the brighter parts are less noisy. Therefore, we want to work on only the fainter parts of the image, otherwise we will introduce unwanted smoothing, or blurring in the brighter parts unnecessarily. It's no surprise that the fainter parts are noisier, because there is less signal in those areas, hence the signal:noise ratio is worse. To improve this, I'm going to use the Multi-resolution Smooth Sharpen routine in ImagePlus.

Open the original image in ImagesPlus (CTRL-O) and then open the Multiresolution Smooth Sharpen control panel by selecting from the main menu Smooth Sharpen - Multiresolution Smooth/Sharpen...

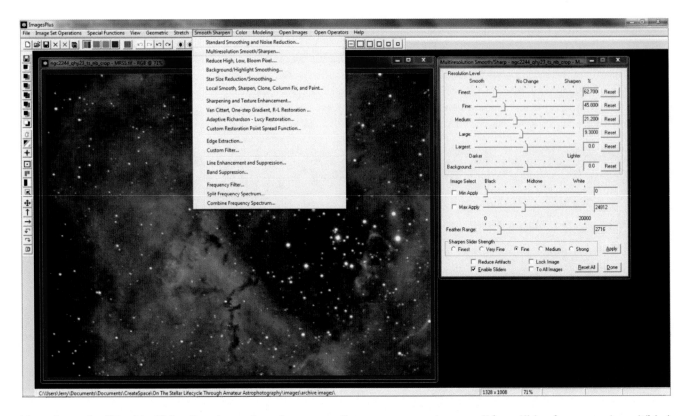

Deactivate the "Enable Sliders" option, otherwise every adjustment you make to a slider will load as a preview. This is useful to do once you've got the basic adjustments set on the sliders so you can see the impact of fine tuning later. The routine acts at different images scales and can be adjusted to work at different levels of aggressiveness. I find that the standard setting of "Fine" is usually sufficient.

Now recall that we only want to smooth the darker parts of the image. So, click on the box next to "Max Apply" and then with the mouse click on the image in an area of brightness that you want to represent the maximum brightness level to apply the routine to. It will work from the min to the max range that you have then set. Again, this can be adjusted live once the Enable Sliders option is turned on later.

Then set the sliders as shown in the image above. We want to be quite aggressive with the "Finest" slider and then progressively less aggressive with the Fine, Medium and Large sliders. And finally, move the "FeatherRange" to about 1000 or so. Now press Apply and you will see the impact of the routine. From this point on, tick the box "Enable Sliders" and play with the various sliders to increase or decrease the strength of the effect and/or apply it to a wider range of image brightness.

Pay particular attention to the areas of the image that you don't think need smoothing, to ensure that they are not getting overly blurred by the smooth routine. If they are, decrease the Max Apply level slider slightly.

The Feather Range slider is very useful, as it allows you to feather the impact of the routine around features that are outside the parameters you've set. So, if there is a bright star in the middle of a part of the nebula that you want to smooth, then you can use the Feather feature to blend the interaction between the bright star and the dark nebula so that you don't end up with harsh borders between bright and dark features.

Side by side below we can see the impact of applying the settings shown above. Perhaps I should have protected the stars a little more with a higher value in the Feather Range. But you can certainly see an improvement in the darker parts of the image in terms of noise reduction. Essentially though this is a blurring routine, so some level of definition is subsequently lost. So be careful how aggressively you apply this routine. Better to do it a little and then if necessary do it a little more, rather than overdo it and lose detail that cannot easily be regained.

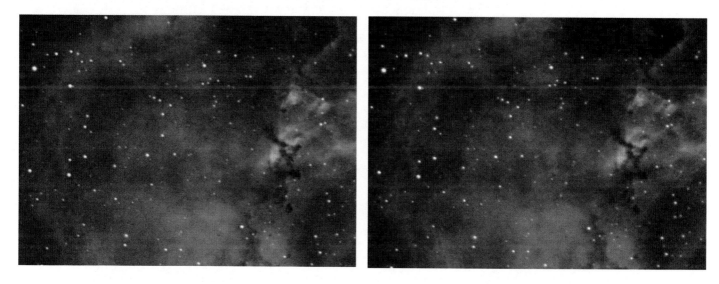

Original image   Smoothed image

The next tutorial will look at enhancing the contrast and detail in parts of the image, recovering perhaps a little of the definition that may have been lost in the smoothing process.

Selectively Enhancing Contrast and Detail

We could use the same MRSS routine in ImagesPlus, reversing the sliders to the right-hand side for sharpening effects rather than smoothing. But I prefer a different solution. One of the most effective ways I've found of *selectively* enhancing contrast and detail in certain portions of an image is to use the High Pass function in Photoshop in conjunction with multiple layers to carefully control the application of the sharpening routine. You can sharpen interesting features in a nebula or galaxy without overly sharpening the stars. We will begin this tutorial with the smooth image we created in the previous tutorial.

Load the smoothed image up in Photoshop, open up the Layers control panel (F7) and press CTRL-J twice to create two copies of the image as separate layers. Your screen will look like this.

The top layer is going to be our control layer. The bottom layer is the original image. And the middle layer is going to be an adjustment layer that applies a High Pass filter with an *overlay* blend mode to sharpen the bottom layer. For now, turn off the visibility of the top layer so you can see what's happening to the lower layers.

Click on the middle layer (Layer 1) and from the main menu select Filter-Other-High Pass. This will open a small control window, which allows you to control the aggressiveness of the filter. Typically, I use the routine in two iterations, one with a small-scale application and one with a larger scale application. So first of all, select a radius of around 6-12 pixels. Note, if you have a very large image (say 20MB or more) you might want to pick a higher radius, and if you have a small image (say only 2MB) you will want to work at small pixel levels. Now change the blend mode from Normal to

Overlay. Your screen should look like this.

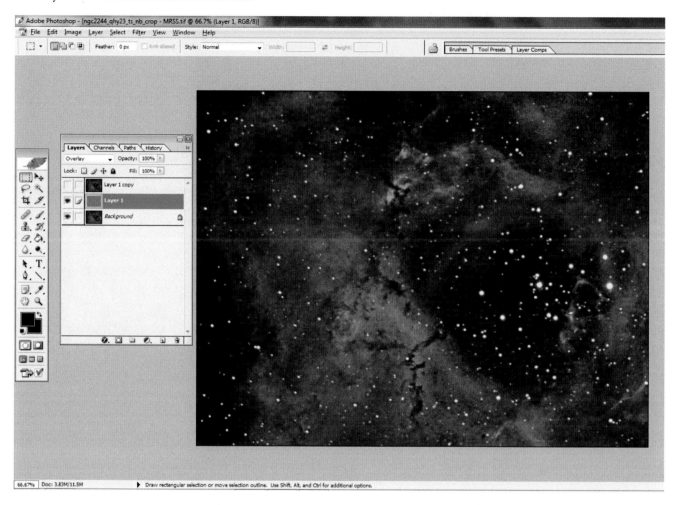

If you toggle the visibility of the middle layer on and off you can see the impact of the sharpening on the original base image. It's probably quite stark and a little too brutal. So, what we want to do now is turn the visibility of the top layer back on and use this as a mask of sorts. We are going to rub out portions of the top layer with the eraser, revealing the sharpened image below. But we're going to do it only in areas where we want to see sharpening applied and we are going to do it in a way that blends the two images seamlessly.

Select the Eraser tool, with an appropriate size for the features that you want to work on. Set the Opacity of the eraser to about 12-15%. (Ensure that you have selected the top layer to work on in the layer control panel.) And begin by erasing some of the more interesting features, such as the dark knotty dust areas in the nebula. Click and hold the left mouse button and move the mouse across the image to erase. If you release the mouse button and click again, rubbing over the same parts of the image you will erase more (another 12% on top of the 12% you already erased). It's a compounding effect. Keep doing this carefully until you have a good balance between the smoother, less noisy parts of the image, and the sharpened interesting features that you want to reveal. Try to avoid stars as you probably don't want to sharpen them much.

Once you're happy with the result, go to the main menu and select Layer-Flatten Image. You may then want to repeat the process with a larger pixel size for the High Pass filter. The unsharpened and the sharpened images are shown below for comparison. Remember, sharpening will introduce noise to the image, so be careful how aggressive you are with it.

Smoothed Image   Sharpened Image

PixInsight

When I first began astroimaging I used to align and stack my images manually with Photoshop. This was a laborious and difficult process and incredibly time consuming. But the advent of sophisticated packages such as ImagesPlus and PixInsight, amongst others of course, has revolutionised the world of astroimaging. I will not go into any detail about how to use PixInsight as there are far better books and tutorials available online. But I thought I would explain some of the routines that I typically use in this software, with some suggestions that I find valuable.

I always use PixInsight now as my default program to do image calibration, alignment and stacking. I find that the results are much better than with anything else I've used. The full list of processes available are shown below, but there are also a number of Scripts that expert users have created that can also be useful shortcuts. One of my favourites is the Script called *DarkStructureEnhance*.

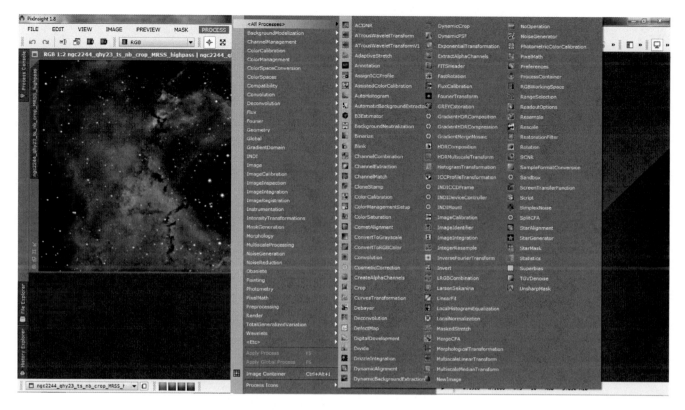

PixInsight calls the image alignment routine *ImageRegistration*. This has some powerful features such as distortion correction and frame adaptation, particularly useful if you're trying to build large mosaic images; and generation of drizzle data to support drizzle integration of the registered files. Stacking the registered frames together is achieved with the *ImageIntegration* tool and the *DrizzleIntegration* tool. The rejection algorithms available in the integration tool are fantastic, really cleaning up the final image, with a wide variety of statistical techniques available to suit the number of images you are stacking.

Other very useful routines are *Background Neutralization* - to automatically balance the colours in the image; *Automatic* or *Dynamic Background Extraction* tools, which can correct for unwanted gradients in the image; *Blink* can be used to rapidly visualise a series of images to check for any trailing or spoiled images before you use them in any processing lists; *Histogram Transformation* is a very useful way of carrying out controlled stretches of the image, with great visualisation/preview features (this is basically the same as the Levels function in Photoshop, but is more sophisticated).

I would highly recommend reading a book edited by the world renowned astrophotographer Robert Gendler called Lessons from the Masters. It gives really important insights into how to process astronomy images, with clear explanations of why activities need to be done. It includes lessons from various amateur imagers around the world, who are all considered experts in their field. Within the book there are some very detailed tutorials on how to get the best from PixInsight - my personal favourite is Bringing Out Faint Large Scale Structures (by Rogelio Bernal Andreo).

Taking Very Long Exposures (Autoguiding)

In the first chapter we looked at the difficulties introduced by taking long exposures with long focal lengths - the stars moving across the field of view very quickly! On page 94 we looked at how to use simple tracking devices, such as the IOptron tracker to counteract this effect, albeit with relatively short focal lengths. So how can we take exposures lasting 5, 10 or even 20 minutes duration with long focal lengths? The answer is *autoguiding*.

Typically, this means having 2 "telescopes" mounted on the same drive, one for imaging and the other for tracking. This isn't always the case, you can use just one telescope and split the light coming out using an off-axis guider like the one shown here. This device uses a small prism to divert some of the light out of the imaging  train and into a separate tracking camera. Some people prefer these as there is no potential for there to be any differential flexure as there can be if two telescopes are mounted together. Personally, I prefer not to use an off-axis guider as it does interfere with the main imaging train to a small degree, and as long as the two telescopes are mounted securely then flexure is unlikely to be an issue.

So how does autoguiding work? The guide telescope is fitted with a small camera, nowhere near as sophisticated or expensive as the main imaging camera, and this takes a series of continuous images of a field of stars and sends the image to your computer. The computer runs a tracking program, such as PHD2 Guiding, and sends signals back to the telescope mount to correct for any small deviations of the guide star from the tracking crosshairs. The principle of autoguiding is illustrated below, using PHD2 Guiding.

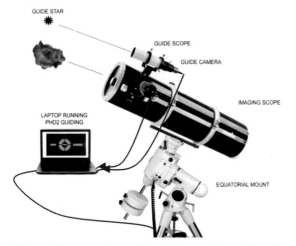

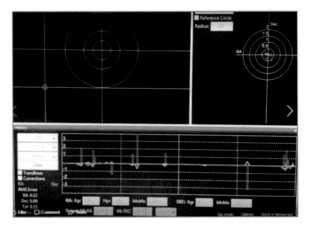

The guide star selected has been registered by the software with a short calibration routine, during which the software instructs the drive mechanism to make certain adjustments. Once it is calibrated, tracking proceeds automatically. The minor adjustments made to the drive can be seen in the blue and red lines. In this case, the adjustments are negligible, at sub-arcsecond levels.

In practice, the accuracy of the tracking mechanism may not be the limiting feature on how much adjustment is required. Things more likely to impact on the accuracy of autoguiding are the level of wind vibration and the amount of turbulence in the atmosphere. With autoguiding it is possible to take extremely long exposures of the object you want to image with your main imaging scope and camera. With my setup, I have three imaging telescopes mounted side by side (TS INED 70mm refractor, Skywatcher MN190 reflector and Takahashi E130 reflector), with a much smaller guide scope (a 50mm straight through finderscope adapted to take the guide camera instead of an eyepiece). The guide camera is a QHY5, which is a relatively sensitive monochrome CCD camera.

Creating The Best Images You Can

This final section contains a few suggestions about how to get the best images you can. You don't have to abide by these principles with every image, but if you do combine all of these suggestions then you'll start to achieve some pretty impressive results. Note, these suggestions are based on taking deep-sky images (nebulae and galaxies).

1. Take the longest exposures that you can. Better to have fewer but longer images than lots of shorter ones. Typically for nebulae I would take 5-minute exposures as a minimum, usually 10 minutes or perhaps 15 or 20 minutes if your guiding system is accurate enough.
2. Take colour images when there is no moon. Otherwise you will get a bright sky background level and the nebula will be washed out. The bright sky will essentially limit the dynamic range available in your image, so you will not get much subtle detail in the nebula.
3. Take narrowband images whenever you like, even if there is a moon, but recognise that you may still need to limit exposure times when there is a moon as there will still be some sky fogging if there is a full moon.
4. Take as many frames as you can to stack together. As an absolute minimum I would suggest 6 frames per filter. But ideally go for 10 or 20 frames if possible.
5. If the number of clear nights you have are limited, you could consider taking colour frames with binning applied on the CCD (to increase the sensitivity), thereby enabling shorter exposures and/or more frames in the same time limit. If you do this, take Luminosity frames at full resolution, or unbinned. The binned colour channels can be combined with the unbinned Luminance data to still give a high-resolution result.
6. If the object emits light in H-alpha, S-ii or O-iii then it's beneficial to take colour (R, G, B) and narrowband data of the object. You can then add narrowband data into the colour channels to enhance the colour data and you can use the narrowband data to create Luminance layers to enhance contrast.
7. If you want high resolution images of large objects you will need to create a mosaic. Overlap the images you take to ensure that there are no seams in the final images. Ensure that the various exposure settings are the same for each frame you take. Make sure that the way you process each component of the mosaic is identical, to avoid differences in brightness levels that are difficult to remove when joining the mosaic together.
8. Exercise patience... I'd aim for one object per imaging session. In the early days of imaging you will be tempted to take images of as many objects as you can in one session. This is ok to start with, but you really need to invest several hours of imaging per object, which in practice means that you can probably only take one object per night - or perhaps one object over several nights!
9. Above all, make sure you enjoy yourself. The hobby can be exasperating at times, but perseverance pays off so don't give up when things get a bit tough.

TO ORDER PRINTS

If you would like to have an unframed print of one of my images, please contact me directly at j.g.hunt@btinternet.com

Prints are professionally done using an Ultra-HD printing process in the UK. A huge range of size choices are available.
Prices typically start from £15 (A6 size); £25 (A3 size); £80 (A0 size) although these are subject to change.
Some images are not suitable to be printed above A4 size depending on the image resolution. Please ask for details.

ABOUT THE AUTHOR

Jeremy Gareth Hunt
CEng (FIMechE), RPP (APM)

I live in the north of England (UK) with my wife, two children and two dogs. I am a Mechanical Engineer by profession, a Project Manager by day and an Astrophotographer by night.

Growing up in the Lake District, the skies were dark and full of stars, but often cloudy. But I never tired of the challenge those clouds posed, snatching as many clear spells as I could to capture the night sky on film. These days, I use digital cameras and maximise imaging time with three telescopes shooting concurrently
from my back-garden observatory.

I have been associated with the Cockermouth Astronomy Society for many years and I try to engage and enthuse new people in astronomy and photography by giving lectures, teaching imaging classes and running an astronomy club at my local primary school. As the billboard slogan goes "pass it on" to the next generation!

Made in United States
Orlando, FL
10 June 2022

18660173R10077